Green Energy and Technology

Climate change, environmental impact and the limited natural resources urge scientific research and novel technical solutions. The monograph series Green Energy and Technology serves as a publishing platform for scientific and technological approaches to "green"—i.e. environmentally friendly and sustainable—technologies. While a focus lies on energy and power supply, it also covers "green" solutions in industrial engineering and engineering design. Green Energy and Technology addresses researchers, advanced students, technical consultants as well as decision makers in industries and politics. Hence, the level of presentation spans from instructional to highly technical.

Indexed in Scopus. **Indexed in Ei Compendex**.

More information about this series at http://www.springer.com/series/8059

Carlos Cuviella-Suárez · David Borge-Diez ·
Antonio Colmenar-Santos

Water and Energy Use in Sanitary-ware Manufacturing

Using Modelling Processes for Water and Energy Accounting and Decarbonisation

Springer

Carlos Cuviella-Suárez
Engineering
ROCA
Barcelona, Spain

Antonio Colmenar-Santos
Control, Electronics and Electrical Engineering
UNED
Madrid, Spain

David Borge-Diez
Electrical Engineering and Systems and Automation Department
University of León
León, Spain

ISSN 1865-3529 ISSN 1865-3537 (electronic)
Green Energy and Technology
ISBN 978-3-030-72490-0 ISBN 978-3-030-72491-7 (eBook)
https://doi.org/10.1007/978-3-030-72491-7

This Springer imprint is published by the registered company Springer Nature Switzerland AG
The registered company address is: Gewerbestrasse 11, 6330 Cham, Switzerland

For my wife Mayte, my daughters, Irina, and Penelope, and my Thesis Directors, Antonio and David who have supported the effort to get this book published.

Carlos Cuviella

For all the doctors that I have directed, who have, undoubtedly, contributed to my academic improvement.

Antonio Colmenar

Thanks to Antonio Colmenar for the opportunity of collaborating in the Ph.D. project and to Carlos Cuviella, for all his commitment in this book: The pupil surpassed the teacher.

David Borge

Preface

One of the main environmental targets of the European Union (EU) is to reduce CO_2 emissions by 40% from 1990 and increase renewable energy by 27% up to 2030. Besides, there is a long-term commitment that by 2050, the EU aims to cut its emissions substantially—by 80–95% compared to 1990 levels as part of the efforts required by developed countries as a group. This commitment includes some key proposals to achieve results like energy efficiency and a deep reform of the emissions trading system (ETS). In the XXI international conference about climate change, held in Paris in late 2015, an international agreement concerning the methods to reduce climate change cutting down total emissions to 45% was signed. This agreement could become a legally binding agreement if it is ratified by at least 50% of members who represent 55% of total emissions.

Other important factors affecting climate change are water consumption, the lack of water and effluent discharges. In the near future, water could become a scarce resource thus many regions might start to experience its supply as a problem. During the last decades, demand for water significantly increased because of many reasons and government as well citizens should take certain actions to overcome the global shortage. The EU has specially raised awareness about this matter, dedicating specific supporting programs for technology development and environmental protection. On April 25, 2014, the Scientific American Review referred to the scarcity of water as a challenge for business development on the headline; "The growing scarcity of freshwater due to rising water demands and a changing climate is increasingly seen as a major risk for the global economy. Consumer awareness, private sector initiatives, governmental regulation and targeted investments are urgently needed to move towards sustainable water use".

Within this framework of energetic and environmental reforms, need arises for countries to start legal development in order to facilitate the fulfillment of their commitment. In this regard, manufacturing companies must approach their modernization in terms of energy or water consumption. It is not just saving but a legal matter.

"Corporate water and energy management" is developing into a very promising means to reduce, energy, water and related costs. Though management is crucial in the current industrial network, some authors consider a wider perspective of the concept including other resources, raw materials and CO_2 emissions reduction. Traditionally, raw materials cost control has been a key point in the cost control in manufacturing. Thus, it is a well-known matter that is not in the scope of this book. From the perspective of this book, the importance when selecting them is the search to minimize energy during the process. Indicators can be applied to quantify the environmental sustainability of raw materials substitution. Embodied energy and CO_2 footprint are proper indicators to evaluate the sustainability of the raw materials. In this book, water is included as a valuable resource to be taken into consideration.

Due to the increasing price of resources and global interest in sustainability, the development of their management is taking an important role as a support function in industrial companies. The impact of them on the final cost is increasingly providing relevance within the cost management. There are some key factors on which the management plan depends. Many new companies addressing resource-related issues have been created during last few years. In fact, ISO released in June 2011 the standard ISO 50001, derived from ISO 9001 referring to energy management systems. The standard aims to help organizations continually reduce their energy use, and therefore their energy costs and their greenhouse gas emissions. In this regard, ISO 24526 aims to water-efficient management systems.

In a conclusion, it can be said that the management and operation of a company´s resources system are of great importance in order to position themselves as leader markets, providing long-term company life throughout sustainability and cost reductions producing, finally an environmental impact reduction. The sanitary-ware industry lacks any research in this field. This book aims to contribute to improve the competitiveness of this industrial sector providing a tool to achieve better performance and sustainability. It provides a practical tool to evaluate the consumptions of water and energy in the field of the sanitary-ware manufacturing to better design new future facilities. However, it does not intend to approach ceramics making. There is a plenty of literature to deal with this subject. It has to be perceived exclusively from the energy, water and CO_2 emission perspectives within a background of sustainability rather than production.

The book has been approached as a handbook or a guide applied to a case study within the sanitary-ware industry. However, the procedure itself might be applied to any kind of manufacturing by just following the appropriate stages adapted to every modeled process.

Barcelona, Spain — Carlos Cuviella-Suárez
León, Spain — David Borge-Diez
Madrid, Spain — Antonio Colmenar-Santos

Acknowledgments The authors want to thank specially SACMI GROUP for all the supporting documentation and technical literature.

Contents

Chapter 1
Introduction to Ceramic Sanitary-Ware Manufacturing

This chapter analyzes the state of the art in the field of sanitary-ware manufacturing so as the global impact of this industrial sector worldwide. Ceramic industry manufacturing requires a great amount of energy and water. Its sustainability and environmental impact demand an effort to develop more efficient technologies to reduce resources consumption. General ceramic industry has been treated from the energy point of view. There is much research regarding tile manufacturing but few articles or papers regarding sanitary-ware from the perspective of the energy–water nexus management. In this regard, the sanitary-ware production sector presents a defined special map of consumption through the manufacturing process characterized by the levels of temperature and heat demands through the different processes where thermal wastes in some of the sub-processes are very suitable to be profited in other ones allowing for an easy reusing of resources.

1.1 Environment, Sustainability and Ceramics

Differences observed in Earth's climate since the early twentieth century evidence that it is changing at a rate that is unprecedented over decades to millennia. Among other natural factors, it seems to be the result of human activity (more than 95% of probability) as NASA states [1].

CO_2 concentration level in the atmosphere is nowadays the highest in 650.000 years. It had never been over 300 ppm before 1950. In 2005, it was 378 ppm and has been growing up to now at a rate of 2.8 ppm/year. On the other hand, temperature is rising 0.2 degrees Celsius per decade since 1950. As a consequence, Artic Sea ice is shrinking 12.8% per decade and ice sheets loss more than 100 Gigatonnes per year. Thus, sea level has risen 178 mm over the past 100 years and it has been growing at a rate of 3.3 mm per year since 1993. These facts provide part of the scientific data needed to understand climate change and evaluating the impact of efforts to

C. Cuviella-Suárez et al., *Water and Energy Use in Sanitary-ware Manufacturing*,
Green Energy and Technology, https://doi.org/10.1007/978-3-030-72491-7_1

combat it. It is up to the global community—the public, policy- and decision-makers and scientific and planning agencies around the world to rule the way to carry it out.

In 1992, the UN adopted the Framework Convention on Climate Change in order to stabilize greenhouse gas concentrations in the atmosphere at a level that would prevent dangerous anthropogenic interference with the climate system. This treaty was extended in Kyoto in 1997. In 2015, the Paris Agreement established the long-term goal to keep the increase in global average temperature to well below 2 °C above pre-industrial levels; and to limit the increase to 1.5 °C, since this would substantially reduce the risks and effects of climate change.

In this background, the European Union (EU) has set itself targets for reducing its greenhouse gas emissions progressively up to 2050 [2]. Key climate and energy targets are set in three stages or packages [3]:

- 2020 climate and energy package. The package sets three key targets:
  - 20% cut in greenhouse gas emissions (from 1990 levels).
  - 20% of EU energy from renewables.
  - 20% improvement in energy efficiency.

  These targets are faced in different areas.

  - Emissions trading system (ETS). This a key tool to meet the target as it covers 45% of the total greenhouse emissions. It has been set to be 21% lower than it was in 2005.
  - National emission reduction targets. They cover 55% of the emissions and differ according to national wealth.
  - Renewable energy—national targets. EU member countries have also taken on binding national targets for raising the share of renewables in their energy consumption by 2020, under the Renewable Energy Directive.
  - Innovation and financing. The EU supports the development of low carbon technologies.
  - Energy efficiency is to be increased by means of a Plan and Directive.
  - Benefits. Meeting the targets, they will be improved the competitiveness, employment security of energy supply and a proper green growth.
- 2030 climate and energy framework. The key targets for 2030 are:
  - At least 40% cuts in greenhouse gas emissions (from 1990 levels).
  - At least 32% share for renewable energy. This value was reviewed from 27%.
  - At least 32.5% improvement in energy efficiency.
- 2050 long-term strategy. The final target is to be climate-neutral area where the green-house emissions are at levels before the industrial development by 1950.

In this regard, the EASME created the IEE in order to create an energy-intelligent future for us all. It supported EU energy efficiency and renewable energy policies, with a view to reaching the EU 2020 targets.

On the other hand, water consumption and effluent discharges are not less important [4, 5]. In a midterm, water could become a scarce good that many regions start to face as a problem in terms of continuity of supply. Water protection is, therefore, one of the priorities of the Commission as stated in the Water Framework Directive. "*European Water Policy should get polluted waters clean again, and ensure clean waters are kept clean*".

The EU has specially raised awareness about this matter dedicating special supporting policies as the [6] and programs for technologies development and environmental protection.

The Blueprint outlines actions that concentrate on better implementation of current water legislation, integration of water policy objectives into other policies, and filling the gaps in particular as regards water quantity and efficiency. The objective is to ensure that a sufficient quantity of good quality water is available for people's needs, the economy and the environment throughout the EU.

The Water Blueprint's time horizon is closely related to the EU's 2020 Strategy and, in particular, to the 2011 Resource Efficiency Roadmap, of which the Blueprint is the water milestone. However, the analysis underpinning the Blueprint covers a longer time span, up to 2050, and is expected to drive EU water policy over the long term.

Though raw materials are not within the scope of this book, but it is important to emphasize that their substitution and proper selection of them can widely influence the consumption of resources [7]. Consequently, CO_2 emissions can vary accordingly. Models have been designed to connect the raw materials and the sustainability [8]. The use of embodied energy and the CO_2 footprint may be very appropriate in accounting for energy and emissions involved in material production.

Within this framework of energy and environmental reforms, need arises for countries to start politics development in order to facilitate the fulfillment of the commitment [9]. In this regard, manufacturing companies must approach their modernization in terms of energy consumption [10]. It is not just an energy saving but a legal matter [11].

"*Energy management*" is a term that is developing into a very promising means to reduce, energy and related costs [12]. Though energy management is crucial in the current industrial network, some authors consider a wider perspective of the concept including other resources and CO_2 emissions reduction. In this book, water is included as a valuable resource to be taken into consideration.

Due to increasing price of energy so as global interest in sustainability developing of resources management is taking an important role as a support function in industrial companies. The impact of the energy in the final cost is increasingly providing relevance to it within the costs management. There are some key factors on which the management plan depends [13]. Many new companies addressing energy-related issues have been created during last years as a result of the pursuing a better efficiency in manufacturing. In fact, ISO released in June of 2011 the standard ISO 50001, derived from ISO 9001 referring to energy management systems. The standard aims to help organizations to continually, reduce their energy use, and therefore their energy costs and their greenhouse gas emissions [14].

In a conclusion, it can be said that a management and operation of a company´s resources system is of great importance in order to position themselves as leader markets, providing long-term company life throughout sustainability and cost reductions producing, finally an environmental impact reduction.

In ceramics manufacturing, energy can account for as much as 30% of the total production cost. Due to the high needs of resources, their price and CO_2 emissions regulation, ceramic producers are facing economical and legal barriers to go on in a sustainable business model. Currently, EU's Horizon 2020 program supports the research, demonstration and market up-take of energy-efficient technologies. CERAMIN has been the particular project aiming the ceramic industry within this program [15].

1.2 Energy and Water Consumption Relevant to the Sanitary-Ware Industrial Sector

Energy is essentially consumed as thermal as well as electrical. Thermal is mainly consumed during kiln firing, drying of intermediates and shaped ware in this order. Currently, the new processes, thermal requirements are based on natural gas [16]. Electricity is consumed throughout the whole sequence of processes to convey, mix, dose, transport, and manipulate material and product. Lighting systems require electricity also. Electrical energy is used in many ways; mechanical energy is used in electrical motors, fans, thermal energy is used to heat the kilns or furnaces, in lighting, etc. Diesel fuel may be required for on-site transportation. Figure 1.1 shows a common energy distribution in a ceramic factory

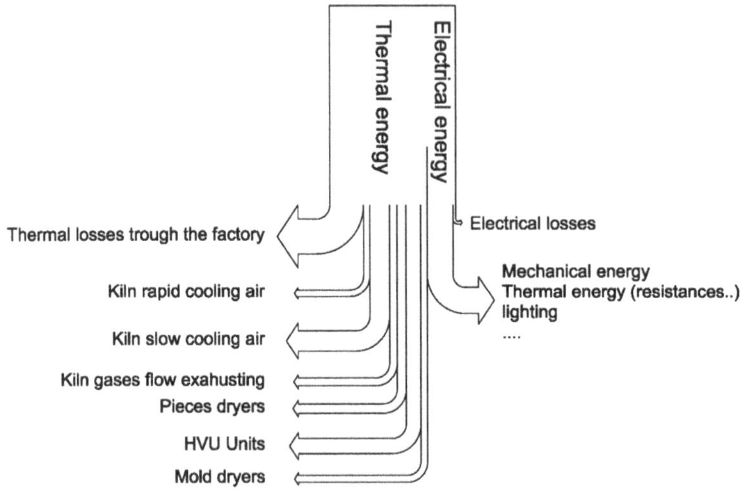

Fig. 1.1 Typical energy distribution in a ceramic factory

Water quality is essential for the preparation of clays and glaze slips. It is used as well as a washing fluid and as a cooling medium. Processed wastewater re-usable techniques take different qualities of water for each process into account.

Despite new technical advances improving the specific consumption of energy, there is still a long way to go up to reach admissible ratios of consumption [17].

Table 1.1 shows the improvement of specific energy consumption in the ceramics industry in Europe.

Assuming that sanitary-ware fixtures have an average weight of over 10 kg, the total energy to produce these pieces is great enough to be taken into consideration. Table 1.2 shows the energy consumption for three sanitary-ware sample factories.

From an environmental point of view, a sanitary-ware factory consumes 130,000 m^3 of water per year to produce about 500,000 pieces [19]. Energy and emission calculations reach up to 360 MJ and 27 kg of CO_2 per piece showing how important the environmental impact of sanitary-ware production can be.

Jingxiang et al. have calculated a life assessment and life cycle costing of sanitary-ware manufacturing in a case study in China finding the following consumption results (Table 1.3).

Table 1.1 Specific energy consumption in the ceramics industry in Europe [18]

Sector	Unit	1980	1985	1990	1995	2000	2003
Bricks and roof tiles	GJ/t	2.65	2.45	2.19	2.06	2.38	2.31
Walls and floor tiles	GJ/t	11.78	9.16	6.76	5.45	5.74	5.60
Refractory products	GJ/t	4.88	4.96	6.51	4.91	5.41	5.57
Sanitary-ware	GJ/t	26.56	24.21	22.27	22.76	20.88	21.87
Vitrified clay pipes	GJ/t			5.75	5.77	6.10	5.23
Table and ornamental ware	GJ/t			47.56	38.91	43.46	45.18
Technical ceramics	GJ/t					34.72	50.39

Table 1.2 Energy consumption data for three simple plants [18]

Parameter	Unit	Plant 1	Plant 2	Plant 3
Production capacity	t/year	10,000	5,120	2,900
Raw material use	t/year	17,000	7,801	3,500
Electrical energy consumption	MJ/kg product	0.36	3.32	3.16
Thermal energy consumption	MJ/kg product	30	22	28

Table 1.3 Energy and water consumption to produce sanitary-ware [27]

	Electricity	Thermal	Water
	MJ/kg product	MJ/kg product	kg/kg product
Consumption	1.56	13.06	15.60

Accordingly with the document on best available techniques in the ceramic manufacturing industry reviewed in August of 2007 [18] at 2.2.4.5., the most extensively used process in the manufacturing of sanitary-ware is the slip casting, either conventional in plaster molds or pressure casting in resin molds.

In this background, this book aims to show the way to decrease the energy requirements in the particular field of the ceramic sanitary-ware manufacturing. Though the tile production has been widely researched from the consumption point of view, no much research has been developed for the sanitary-ware. It has to be remarked that both are ceramic processes with almost the main manufacturing global method, although some important differences make the latter particularly interesting for the resources reduction study.

This book approaches these particularities to re-engineer the map of single energy and water flows reducing the global consumption in a factory type [21]. In this way, it provides a calculated way to apply the scheme of resource consumption to quantify the global savings of both, energy and water.

Looking for research in the field of resource management in manufacturing, it was found that no research in this field had been carried out related to sanitary-ware production. May et al. have widely studied this subject over 365 articles published from 1995 to 2015 [22]. They concluded that the integration of energy efficiency in manufacturing has been recognized as a means to foster economic and environmental performance, increase competitiveness and spur innovation. As stated in the introduction, energy and resource consumption is a major focus for industry, policymakers, and society with manufacturing units and factories playing a critical role to achieve these targets but there is a lack of research in this regard for the sanitary-ware industry and this is the main aim of this book.

On the other hand, state of art sanitary-ware manufacturing has been studied in detail in the [18] (BAT).

1.3 Sanitary-Ware Manufacturing Worldwide

Currently, in Europe, the ceramic industry employs 338,000 persons and provides a business turnover of 27.8 billion euros [23]. In 2008, the European ceramic business reached a value of 28 billion euros and the world ceramics market is worth in the region of €120 billion [24]. The figures have barely changed.

In 2002, the European sanitary-ware industry turnover was 2 billion euros [25]. Nowadays, it is about 1.67 billion euros [26] and represents 9% of the ceramic employment.

The production share is led by five main production countries Fig. 1.2.

As per ACIMAC [27], the European Union produced 44.6 million pieces in 2004 and 41.5 in 2014. Although the crisis has reduced the pace of production in Europe, it is still highly relevant. The research shows that 216.6 million pieces were globally produced in 2004, rising to 349.3 million in 2014.

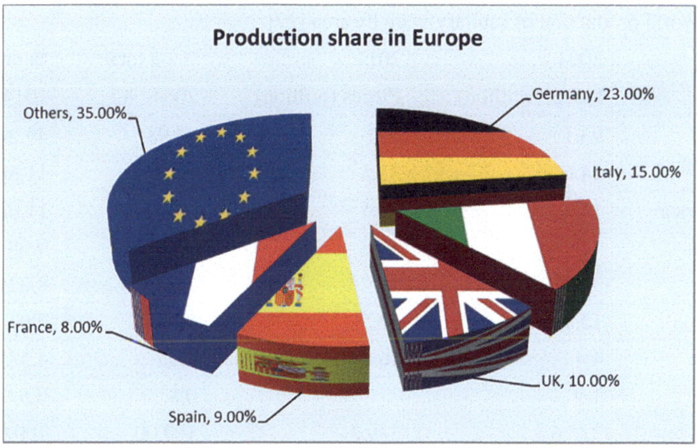

Fig. 1.2 Production share of sanitary-ware in Europe. *Source* Competitiveness of the Ceramics Sector [24]

World sanitary-ware production increased 61.3% from 2004 to 2014, at a growing rate of 4.9%.

Asia has been the largest production area. At current, it represents about 50% of global production. Almost 70% of this share corresponds to China, which is the largest producer worldwide. Nevertheless, the greater growths correspond to Africa, Middle East and South America, where Brazil is the most relevant producer.

These figures justify the negative ones for EU and NAFTA due to the globalization and production area displacements. Pay attention to NAFTA decrease even the relevant growth of Mexico, which has almost tripled its production.

On the other hand, BRIC and Next 11 areas have relatively important growth rates over the modest average of 27% in other countries.

World sanitary-ware production is also increasingly concentrated in the hands of a few very large multinational groups. It is estimated that currently, the top 20 groups (excluding those in China) produce over 43% of the total world production (Table 1.4 and Fig. 1.3).

In this scenario of industrial production for the sanitary-ware sector and assuming an average of 10 kg per piece, it can be derived from Table 1.3 that about 54 Mm^3 of water, 51 PJ are needed worldwide to produce sanitary-ware and 8.36 Mt of CO_2 are delivered. These figures should draw anyone's attention to this industry sector, which, in addition, is supposed to be so sensitive to energy price impact.

Ceramic industry has been extensively studied as a general subject [28]. However, no much research is devoted specifically to sanitary-ware manufacturing. This industrial sector is particularly interesting from the thermal point of view. The great thermal power required to manufacture sanitary-ware articles is required in different sub-processes of the factory at different thermal levels. As a matter of fact, the temperature and amount of residual heat from firing are highly suited to supply other heating processes as drying or heating [29]. It can be clearly deduced the way to

Table 1.4 World production of sanitary-ware by area [27]

Area	2004	2014	% on total	% on total
	Pieces (million)	Pieces (million)	2004 (%)	2014 (%)
Asia	93.1	172	42.98	49.24
EU	44.6	41.5	20.5	11.88
South America	14.6	38.5	6.74	11.02
Middle East	12	33.8	5.54	9.68
NAFTA	33.9	28	15.65	8.02
Other Europe	12.1	17.5	5.59	5.01
Africa	4.4	16	2.03	4.58
Oceania	1.9	2	0.8	0.57
World	216.6	349.3	100.00	100.00

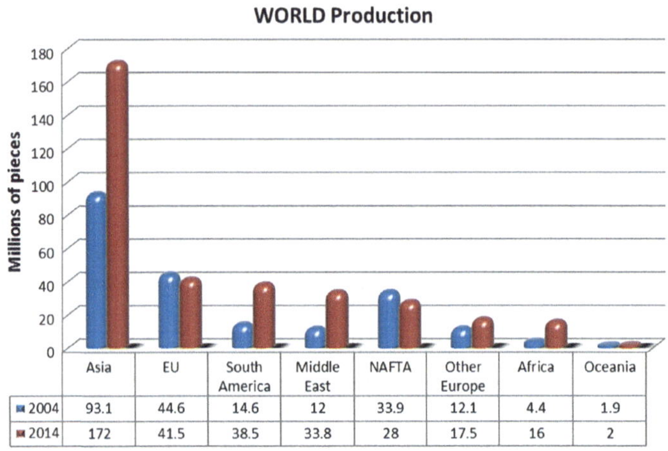

Fig. 1.3 World Sanitary-ware production by area

apply thermal engineering to save up to one third of the energy. As suggested in [30], thermal requirements are demanded at different levels within the EU region. In 2012, 1035 TWh were demanded over 500 °C, 706 TWh in the range of 100–500 °C and 228 TWh below 100 °C.

1.4 Ceramic Sanitary-Ware Manufacturing General Process

The general process used in the ceramic industry for sanitary-ware consists of different sub-processes or basic industrial operations as shown in Fig. 1.4.

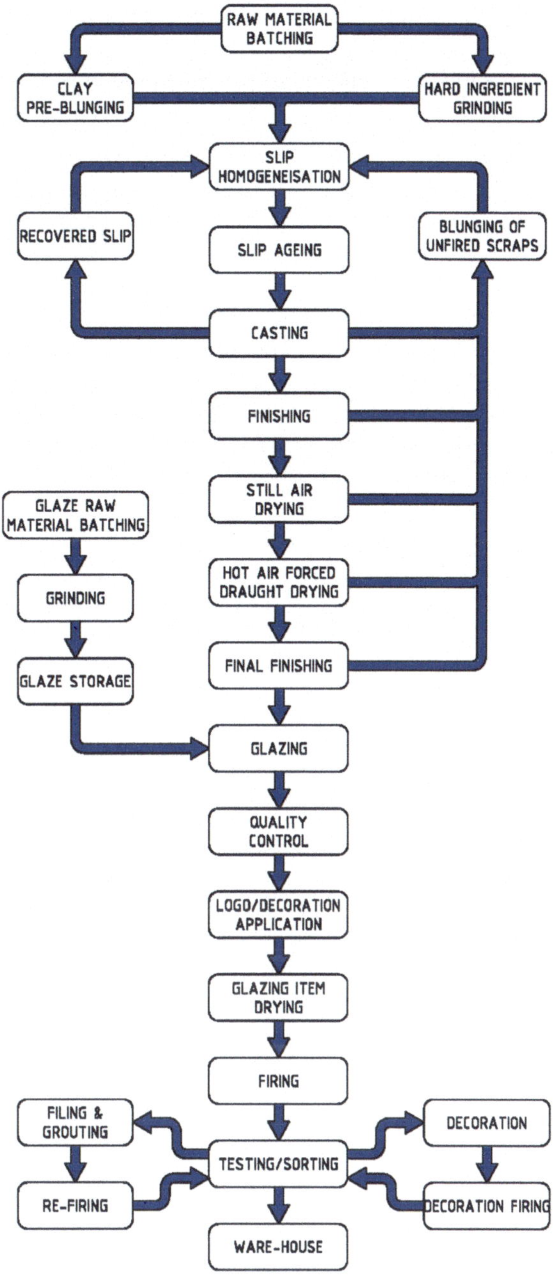

Fig. 1.4 General process of sanitary-ware manufacturing. *Source* Ceramic Technology Sanitary Ware SACMI 2010 [31]

Ceramic production is not the scope of this book but the energy and water requirement analysis so as their reduction. In this regard, a shallow explanation of the manufacturing system and its different processes is defined as follows [32]:

- Non-plastic raw materials are milled and mixed with plastic material in a blunger where they are homogenized in aqueous medium. Sometimes the required quality of the water demands salt removing treatment by R.O.
- The resulting barbotine is cast into either plaster or the resin molds where part of the water is eliminated by filtration through the porous resin. This mold has to be flushed with warm and salts free water to get it cleaned. Usually, this water is provided after a R.O. treatment to the boilers where it is warm up.
- Molded piece is taken out and left within the casting hall to set the remaining humidity before it can be taken to the dryer. The hygrothermal environment has to be kept within stated range by means of HVU's.
- After that, the ware is dried in the dryers where hygrothermal environment is again kept within stated range. Dryers have to take off the humidity of the ware before it is taken into the kiln.
- Finally, ware is put into the kiln to be fired after spraying a thin lay of glaze. The purpose of the kiln is to rise the temperature of ware up to 1,523 K in order to produce the sintering of the ceramics and vitrification of the glaze. Later on, the ware has to be cooled in a controlled manner to allow for the chemical processes within the ceramics. Ware firing uses to be an exothermic process.

In sanitary-ware industry, the ware has to be dried after casting and before firing on the contrary of other ceramic processes under controlled conditions. Besides, the hygrothermal environment in the casting hall has to be kept into a defined range. On the other hand, when resin molds are used to cast, much treated water is wasted to flush the mold in every casting cycle [32].

The general process is essentially the same throughout different technologies. The only remarkable difference that can be found regards the casting. In particular, the material the molds are made of. Either plaster or resins are the two common filtrating media to cast the slip or barbotine.

The general process for the two most extended technologies; associated consumptions and the global impact for this particular industrial sector are described in Chap. 2.

References

1. NASA Climate Change (2019) Climate change: vital signs of the planet, 28 May 2019. https://climate.nasa.gov/. Accessed 12 June 2019
2. European Commission (2016) PRESS RELEASES - Press release - Sustainable Development: EU sets out its priorities, 22 November 2016. http://europa.eu/rapid/press-release_IP-16-3883_en.htm. Accessed 12 June 2019

3. European Commission (2018) PRESS RELEASES - Press release - 2030 climate and energy goals for a competitive, secure and low-carbon EU economy, 19 February 2018. http://europa.eu/rapid/press-release_IP-14-54_en.htm. Accessed 13 June 2019
4. Drought and water overuse in Europe (2009) European Environment Agency, 12 February 2009. https://www.eea.europa.eu/media/newsreleases/drought-and-water-overuse-in-europe. Accessed 11 June 2019
5. Water Scarcity & Droughts in the European Union (2016) Scarcity and droughts, 08 June 2016. http://ec.europa.eu/environment/water/quantity/scarcity_en.htm. Accessed 11 June 2019
6. European Commission (2012) A blueprint to safeguard Europe's water resources, 14 November 2012. https://eur-lex.europa.eu/legal-content/EN/TXT/?uri=CELEX:52012DC0673. Accessed 13 June 2019
7. Mezquita A, Monfort E, Ferrer S, Gabaldón-Estevan D (2017) How to reduce energy and water consumption in the preparation of raw materials for ceramic tile manufacturing: dry versus wet route. J Clean Prod 168:1566–1570. https://doi.org/10.1016/j.jclepro.2017.04.082
8. Bontempi E (2017) A new approach for evaluating the sustainability of raw materials substitution based on embodied energy and the CO2 footprint. J Clean Prod 162:162–169. https://doi.org/10.1016/j.jclepro.2017.06.028
9. Thollander P, Kimura O, Wakabayashi M, Rohdin P (2015) A review of industrial energy and climate policies in Japan and Sweden with emphasis towards SMEs. Renew Sustain Energy Rev 50:504–512. https://doi.org/10.1016/j.rser.2015.04.102
10. Bonacina F, Corsini A, De Propris L, Marchegiani A, Mori F (2015) Industrial energy management systems in Italy: state of the art and perspective. Energy Proc. 82:562–569. https://doi.org/10.1016/j.egypro.2015.11.871
11. Bigerna S, Bollino CA, Micheli S (2016) Renewable energy scenarios for costs reductions in the European Union. Renew Energy 96, Part A:80–90. https://doi.org/10.1016/j.renene.2016.04.024
12. Schulze M, Nehler H, Ottosson M, Thollander P (2016) Energy management in industry – a systematic review of previous findings and an integrative conceptual framework. J Clean Prod 112, Part 5:3692–3708. https://doi.org/10.1016/j.jclepro.2015.06.060
13. Sa A, Thollander P, Cagno E (2017) Assessing the driving factors for energy management program adoption. Renew Sustain Energy Rev 74:538–547. https://doi.org/10.1016/j.rser.2017.02.061
14. Giacone E, Mancò S (2012) Energy efficiency measurement in industrial processes. Energy 38(1):331–345. https://doi.org/10.1016/j.energy.2011.11.054
15. CERAMIN PROJECT (2009) Energy saving concepts for the European ceramic industry. Intelligent Energy Europe, 31 October 2009. https://ec.europa.eu/energy/intelligent/projects/en/projects/ceramin. Accessed 02 Aug 2018
16. Akinbode FO (1995) Energy utilization in the clay and pottery processing industries in Nigeria. Energy 20(6):585–587. https://doi.org/10.1016/0360-5442(94)00086-I
17. Weisser D (2007) A guide to life-cycle greenhouse gas (GHG) emissions from electric supply technologies. Energy 32(9):1543–1559. https://doi.org/10.1016/j.energy.2007.01.008
18. Reference document on best available techniques in the ceramic manufacturing industry. European Commission, August 2007 (2007). http://eippcb.jrc.ec.europa.eu/reference/BREF/cer_bref_0807.pdf. Accessed 02 Dec 2016
19. LIFE_ENVIP.pdf (2012). http://ec.europa.eu/environment/life/project/Projects/index.cfm?fuseaction=search.dspPage&n_proj_id=4630&docType=pdf. Accessed 02 Dec 2016
20. Lv J, Gu F, Zhang W, Guo J (2019) Life cycle assessment and life cycle costing of sanitary ware manufacturing: a case study in China. J Clean Prod 238: https://doi.org/10.1016/j.jclepro.2019.117938
21. Cuviella-Suárez C, Colmenar-Santos A, Borge-Diez D, López-Rey Á (2018) Management tool to optimize energy and water consumption in the sanitary-ware industry. J Clean Prod 197, Part 1:280–296. https://doi.org/10.1016/j.jclepro.2018.06.195
22. May G, Stahl B, Taisch M, Kiritsis D (2017) Energy management in manufacturing: From literature review to a conceptual framework. J Clean Prod 1–26

23. Ceramics - Growth - European Commission Growth, 20 January 2017 (2017). /growth/sectors/raw-materials/industries/non-metals/ceramics_en. Accessed 02 Dec 2016
24. Growth - European Commission - Publications 2008, 13 October 2008. http://ec.europa.eu/growth/tools-databases/newsroom/cf/itemdetail.cfm?item_id=3440&lang=en&title=FWC-Sector-Competitiveness-Studies—Competitiveness-of-the-Ceramics-Sector. Accessed 02 Dec 2016
25. Bulletin of Ceramic Spanish Society, 02 March 2004. http://boletines.secv.es/upload/2007013192112.contenido_adicional.pdf. Accessed 02 Dec 2016
26. Ceramic-industry-roadmap-paving-the-way-to-2050.pdf, 27 November 2012. http://cerameunie.eu/topics/cerame-unie-sectors/cerame-unie/ceramic-industry-roadmap-paving-the-way-to-2050/?media=4249&f=Ceramic%20Roadmap%20to%202050%20EN.pdf. Accessed 02 Dec 2016
27. Ceramic World Review 114/2015, 18 December 2015. http://www.ceramicworldweb.it/cww-it/magazines/ceramic-world-review-1142015/. Accessed 02 Dec 2016
28. Ciacco EFS, Rocha JR, Coutinho AR (2017) The energy consumption in the ceramic tile industry in Brazil. Appl Therm Eng 113:1283–1289. https://doi.org/10.1016/j.applthermaleng.2016.11.068
29. Christos A, Theocharis T (2001) Energy saving technologies in the European ceramic sector: a systematic review - ScienceDirect. Appl Therm Eng 21(12):1231–1249. https://doi.org/10.1016/S1359-4311(01)00006-0
30. Rehfeldt M, Fleiter T, Toro F (2018) A bottom-up estimation of the heating and cooling demand in European industry. Energy Effic 11(5):1057–1082. https://doi.org/10.1007/s12053-017-9571-y
31. Ceramic Technology Sanitary-ware, 2010th edn. SACMI IMOLA S.C. (2010)
32. King AG (2001) Ceramic technology and processing, 1st edn. Andrew, William. Elsevier, xxx

Chapter 2
Types of Factories by Casting Technology

When starting a proper plan to manage resource consumptions, it has to be understood the global process from the correct perspective. The production line may appear quite different depending on the observer. When observed from consumption point of view, parameters and variables are not the same as from a financial or production perspective. In this regard, the picture of the factory has to be drawn with a proper plot of consumption points. Quantities, the way they are required, wastes or leftovers, mass and energy flows, etc. are magnitudes to be precisely taken into account. Thus, for the purpose of this book, factory has been focused as a sequence of water and energy points of consumers and either, existing or potential links between them.

2.1 Main Technologies for Casting Barbotine

There are two main different ways to cast barbotine into the mold in order to obtain a relative consistent piece [1].

2.1.1 Traditional Technology

It consists of plaster molds where the barbotine is cast. The plaster takes out the water from the barbotine reducing the humidity levels down to 22–24% after the casting time, which can vary from some minutes up to one hour depending on the piece size. When this time is reached the mold is opened, and the piece removed at the desired humidity. This is a cyclical process where the mold has to release the water absorbed making this step of the process the weakest point of this technology since depending on the way the mold is dried, its capability to remove water again will vary making it inefficient for the next cycle. If the drying is quick, the capillarity stream could be cut, the water will not be able to reach the surface of the plaster mold

C. Cuviella-Suárez et al., *Water and Energy Use in Sanitary-ware Manufacturing*, Green Energy and Technology, https://doi.org/10.1007/978-3-030-72491-7_2

Table 2.1 Environmental parameters in casting hall (*Source* [2])

	Temperature	Humidity	Water flushing
	°C	%	dm^3/(mold·cast)
Plaster	40	50	5
Resin	25	75	180

and the mold will not be dried properly and the next cycle will not run efficiently. On the other hand, if the drying is too slow, the mold will not be dried enough for the next cycle and the water absorption will not reach the desired level. The plaster mold drying is consequently, a critical limiting factor to get high productivity, since one mold is able to produce about two pieces per day. In addition, the psychrometric environment of the casting hall has to be maintained within specific temperature and moisture conditions in order to achieve the proper drying cycle. The main differences are shown in Table 2.1.

2.1.2 *Current Technology*

Currently, an alternative technology has been developed in order to increase productivity. When the plaster mold is replaced by a resin mold, the productivity is highly increased since this mold is able to release the capillary water by just blowing it. Thus, the productivity of one resin mold can reach up to 80 pieces per day depending on the size. The weak point of this technology is the water waste produced during the mold washing. For every cycle, the mold has to be washed by means of allowing clear water to flow through it, so all the salt retained along the capillaries is removed to preventing salts building-up, which would end up blocking the capillarity suction. The water needed to wash the mold can reach 180 dm^3 per mold cycle depending on the mold size. In addition, this water has to be free of salts, so it has to be previously treated by either, reverse osmosis or distillation.

The research is focused on the sanitary-ware as a particular industrial area. The general process will be briefly described initially, then different processes will be analyzed and finally they will be evaluated from an energetic point of view [3]. As sanitary-ware is a great consumer of heat and water in several stages of production, the consumption will be calculated for each step and then study the way to link heat wastes from one process to another with lower heat requirement in order to reduce the global need of raw energy and the way to reduce the water consumption. A comparable research has been made related to brick making in China [4].

Based on the main differences of the global manufacturing process, two study cases will be taken into consideration as most of the factories can be matched to one or the other.

As an example, Table 2.2 shows the different resource balances for both technologies, plaster or resin molds casting.

Table 2.2 Energy and water consumption for the two main technologies (*Source* author)

Nº Pieces	500,000 pcs/year			
Consumptions	Electrical energy	Thermal energy	Water	Water
	GJ/year	GJ/year	m^3/yr	t_{CO_2}/yr
Plaster casting	27,609	85,531	32,930	9,420
Pressure casting	22,599	84,535	242,496	8,542

Conventional sanitary-ware production uses plaster molds, which operate on the principle of capillarity. The plaster works like a sponge and absorbs the water from the slip. Plaster molds operate at low pressure and are relatively inexpensive, but they do carry several drawbacks. Forming and reuse times are quite lengthy, and the molds typically have a limited working life, often making them impractical for large-volume operations.

Porous resins have been used as an alternative mold-making material for several decades, and their use has coincided with the introduction of higher and higher casting pressures. Today, pressures of 1.3–1.5 MPa are common. The resins work like filters, trapping the ceramic body particles contained in the slip while drawing off the water. Pressure casting with resin molds has resulted in shorter casting cycle times, more compact casting departments and significantly reduced labor requirements. However, the high pressures and complexity of these automatically controlled cycles have led to the development of machines that require a considerable economic investment. Hence, their use can only be justified where the output is particularly high and the product mix not too demanding.

In most cases, the resins themselves dictate the use of high pressures. Two types of resins are used for pressure casting: microporous and macro porous. The former has an average pore diameter of no greater than 10 μ, while the latter normally have average pore diameters of around 20 μ. However, both resin types have a very low, or non-existent (in the case of macro porous resins) thickness-forming capacity, i.e., a very low capacity to remove the water contained in the slip, as shown in Table 2.3.

In the last column of the table, the thickness-forming capacity (at atmospheric pressure) for various resins is compared to that for plaster over a period of 60 min. The thickness formed by the plaster is defined as 100 and acts as the reference number for values assigned to the other materials. A look at the table shows that the thickness-forming capacity of microporous resins is 30–40% of plaster capacity, while macro porous resins fail to reach even the 10% mark. Consequently, if a sanitary-ware article is produced using a resin mold, it is absolutely necessary to apply a certain pressure (from 0.6 to 0.7 MPa upwards) to the slip. At lower pressures, formation times will be unacceptably long.

Pressure casting techniques have now reached the levels of reliability and quality that make their application imperative where manufacturing conditions, output volumes and the product mix permit. When these conditions are unattainable or have been only partially achieved the pressure casting technology does not provide proper yields which justify the great initial investment. This problem is particularly

Table 2.3 Comparison of typical porous resins used in the sanitary-ware industry (*Source* SACMI resins specifications)

Porous resin type	Working pressure	Avg. Pore diameter	Open porosity	Outflow rate[a]	Thickness forming power (atmospheric pressure)[b]
	10^5 Pas	μm	%	dm^3/min	60 min
Low pressure	0–1	0.5	45–50	0	120–130%
		1.0	45–50	0	100–110%
		1.5	45–50	0.17	75–85%
Microporous resin 1 (SACMI-POR)	6–13	5–7	27–37	25	40%
Microporous resin 2 (Other)	13	5–10	27–28	33.3	30%
Macro porous resin 1 (SACMI-POR)	13	20–25	28–29	250	<7%
Macro porous resin 2 (Other)	13	20–26	30–32	250	<7%

a—Refers to a resin thickness of 2 cm with filtration cross-section of 40 cm^2 at a dynamic water pressure of 4•105 Pa
b—Calculated by casting at atmospheric pressure and comparing data with plaster performance (which was assigned a value of 100)

evident in the production of toilets and bidets, complex articles that can only be produced in four or five-part molds. Such manufacturers often fail to achieve the output volume and production mix thresholds that justify the use of resin molds.

Matsumoto states in [5] some of the various benefits of using pressure against conventional casting. On the other hand, resin molds are quite intensive consumers of high-quality water. Since they act as a filtering media, during each cycle they have to be flushed with clear and salts free water in order to leave them ready to cast a new piece. The water leaving the mold during the casting process goes through the filtering channels dragging solid particle so as dissolved salts that are decanted along the porous-media. After cycle is finished, the clear water has to be pumped backwards in order to erase those particles and salts. Otherwise, the filtering media efficiency would drop drastically.

2.2 Characterization of Factory Type

Based on the stated technologies, two different types of factory are going to be modeled where only processes related to mold making and casting differ each other. In order to compare them, a sole structure of the factory will be applied to both types,

Table 2.4 Characterization of the factory by production sections. (*Source* Data compiled by the authors)

	Section	Process
01	Raw materials	Raw material stock and conveying
02	Slip preparation	Milling of hard materials and blunging of soft materials, so as dosing proper formula for slip
03	Glaze preparation	Milling of hard materials and blunging of soft materials, so as dosing proper formula for glaze
04	Modeling and molds	Production of matrixes for molds and molds, either from plaster or resin, so as to plaster mold drying
05	Casting	Casting of slip into molds
06	Heating and ventilation	Temperature and moist control of the casting hall
07	Drying	Ware drying by controlled air heating ventilation
08	Glazing	Application of glaze to ware by humid spraying
09	Kilns	Ware firing for sintering of ceramics in a first fire and occasionally, second fire upon minor failures of first fire
10	Sorting	Quality control and sorting. Packaging
11	Offices	Administrative section
12	Ware-house	General store and final product ware-house
13	Facilities	Facilities to operate the factory: compressed air, boiling, electrical, pumping, etc.

based on the various sections along the manufacturing line according to flow chart of the global process in Chap. 1. Consumption differences may appear upon these sections depending on the processes and corresponding equipment (Table 2.4).

Section by section, consumptions will be analyzed as per process needs and applied equipment along the production duty cycles. Results allow to assess potential improvements and quantifying them. To compute the analyses of consumptions, mathematical models will be designed for every required process as per common available equipment.

2.2.1 *Water Consumption Differences*

According to main technologies description, main water consumption difference is directly linked to mold making and casting processes. On the other hand, it has been stated that water has to be provided salt-free. The difference of water consumption between technologies is shown in section where it has been split up by sections.

Plaster molds making involves a relevant amount of water to cast plaster whereas resin molds do not require water within the production process. On the other hand, for the slip casting process, resin molds require up to 0.2 m^3 of treated water to flush them after every casting cycle. Considering that this water is obtained by reverse osmosis, the total primary amount for this purpose is about double for a rough efficiency of 0.5 [6].

The international center for sustainable development of energy, water and environment systems (SDEWES) holds a series of conferences with the target of improving efficiencies by integrating various life-supporting systems, using waste from one, as resource in other, and in exact moment when it is beneficial to all, integrating electricity, heating, cooling, transport, water, buildings, waste, wastewater, industry, forestry and agriculture systems [7].

This is the key point of this book from the water-saving perspective. Substituting the current reverse osmosis water treatment by distillation technology, the water can be reused and the wasted heat from the kiln is profited as energetic resource for water treatment. There is a great potential in heat recovery to improve energy efficiency [8] or economic aspects [9] with no dependence on the temperature level [10], so this paper provides a specific application of that general approach.

2.2.1.1 Reverse Osmosis

Currently, there are many applications of reverse osmosis within different industrial areas with the only purpose to treat water in order to reduce salt contents which affect either efficiency of the process or the life cycle of the machines [11].

This is the case of the sanitary-ware industry where both effects are present in the molds' maintenance and the efficiency of the casting process. Manufacturers of HPCM's recommend flushing them after every cycle with clean and warm water. The usual way in factories today, to produce this water is R.O. By R.O., the water produced is suitable for production as well as for flushing the molds. The point is that flushing water is rejected after cleaning the mold carrying dissolved salts so as slip residual particles. This water is, under these circumstances, useless for a new washing cycle. Membranes into the R.O. device are quickly clogged when suspended solids reach them. For this reason, it is not advisable to reuse water from washing within a R.O. system.

2.2.2 Energy Consumption Differences

2.2.2.1 Electrical Differences

The electrical differences lie mainly in casting section. Mold making process is essentially the same from the electrical point of view. Mixing components, casting and drying or curing, there may be a slight difference related to dryer for plaster molds. There are ventilation and exhausting fans to remove water from plaster. On the other hand, for resin molds, it is rather a curing process than a proper drying process. At the facilities section, it has to be allowed for a R.O. plant in a resin molds factory. Pumps for this plant represent another electrical difference.

Regarding casting process, it is remarkable the different size of area for both technologies. Required area of casting section per mold is quite similar for both technologies. On the contrary, the ratio of production varies up to 30 depending on the duty cycle. The area ratio is inversely proportional.

While resin technology requires much water to flush molds, plaster requires controlled environment to dry molds by means of heat ventilation units provided with powerful fans.

2.2.2.2 Thermal Differences

Mold dryers in plaster mold factories are heat consumers that do not exist in resin mold factories.

Regarding casting areas, hygrometric conditions are required to prevent ware against cracks. Hence, air heating is not a thermal difference because of ware. But the real and important difference lies in drying plaster molds. When cast, the barbotine leaves more than 50% of humidity within the mold capillaries. This water has to be evaporated later in order to let a new casting cycle by blowing extra heated air during the drying period of casting hall.

Water to flush molds is required to be at about 45°C. Conventional boilers are usually provided for this purpose. Due to high flushing water demand, this is another important thermal difference.

References

1. Reference Document on Best Available Techniques in the Ceramic Manufacturing Industry. European Commission, August 2007 (2007). http://eippcb.jrc.ec.europa.eu/reference/BREF/cer_bref_0807.pdf. Accessed 02 Dec 2016
2. Ceramic Technology Sanitary-ware, 2010th edn. SACMI IMOLA S.C. (2010)
3. Petek J, Glavič P, Kostevšek A (2016) Comprehensive approach to increase energy efficiency based on versatile industrial practices. J Clean Prod 112, Part 4:2813–2821. https://doi.org/10.1016/j.jclepro.2015.10.046
4. Zhang Z (1997) Energy efficiency and environmental pollution of brickmaking in China. Energy 22(1):33–42. https://doi.org/10.1016/S0360-5442(96)00078-3
5. Matsumoto A Slip-casting by water-absorbing resin mold enables crack-free CeRamic MoldiNg system and products with partially different thicknesses. In: Ohji T, Singh M, Halbig M (eds) Advanced processing and manufacturing technologies for nanostructured and multifunctional materials II. John Wiley & Sons, Inc., pp 45–54 (2015)
6. Fritzmann C, Löwenberg J, Wintgens T, Melin T (2007) State-of-the-art of reverse osmosis desalination. Desalination 216(1):1–76. https://doi.org/10.1016/j.desal.2006.12.009
7. Urbaniec K, Mikulčić H, Duić N, Lozano R (2016) SDEWES 2014 – sustainable development of energy, water and environment systems. J Clean Prod 130:1–11. https://doi.org/10.1016/j.jclepro.2016.04.062
8. Miró L, Brueckner S, McKenna R, Cabeza LF (2016) Methodologies to estimate industrial waste heat potential by transferring key figures: a case study for Spain. Appl Energy 169:866–873. https://doi.org/10.1016/j.apenergy.2016.02.089

9. Brückner S, Liu S, Miró L, Radspieler M, Cabeza LF, Lävemann E (2015) Industrial waste heat recovery technologies: An economic analysis of heat transformation technologies. Appl Energy 151:157–167. https://doi.org/10.1016/j.apenergy.2015.01.147
10. Taylor PB, Wadkin JW (1986) Industrial heat recovery projects supported through the energy efficiency demonstration scheme. J Heat Recover Syst 6(1):83–92. https://doi.org/10.1016/0198-7593(86)90175-X
11. Colla V, Branca TA, Rosito F, Lucca C, Vivas BP, Delmiro VM (2016) Sustainable reverse osmosis application for wastewater treatment in the steel industry. J Clean Prod 130:103–115. https://doi.org/10.1016/j.jclepro.2015.09.025

Chapter 3
Planning for Energy and Water Management

Once the main manufacturing line has been defined in the previous chapter by describing the two main process sequences, it is crucial to state the individual steps through them. Prior to arrange any action to improve the performance of the system, it has to be analyzed in detail for every individual step without losing sight global links between them. Only with a proper perspective of the general picture, it is possible to understand the spots where the resources are consumed, the level of impact they represent overall needs of the plant. In this regard, this chapter aims to describe how to arrange data collection and the way to quantify the variables that affect the global performance by addition of partial systems. Physical and thermodynamic models are developed to assess energy and water requirements so as corresponding wastes. Modeling processes procedure and thermal engineering have been applied to show the potential of saving and to identify and quantify the equipment more likely to reduce energy [1].

3.1 Characterization of Consumption and Waste

From the electrical point of view, a sanitary-ware factory has not got any particular characteristic. Electrical requirements are quite like any other type of factory. On the other hand, configuration of energy heat requirements and temperature levels through the different sub-processes in the manufacturing sequence are distributed in such a way that a proper network of recovered heat can save up to 1/3 of the primary energy consumption of the factory. A half of the total thermal requirement is devoted to fire the ware along a tunnel kiln that exhausts 3/4 of this heat through the exhausting and cooling stacks at a temperature level to be recovered in the rest of common thermal processes as drying or heating. Besides, the great water consumption involved in ceramic manufacturing might be recycled by proper treatment fueled with waste recovered heat. Due to staggered use of heat, there is also a chance to couple a combined heat and power plant, which reduces the use of primary energy.

C. Cuviella-Suárez et al., *Water and Energy Use in Sanitary-ware Manufacturing*, Green Energy and Technology, https://doi.org/10.1007/978-3-030-72491-7_3

The starting point of the management planning is to draw the energy map of the factory to set the energy consumption pattern [2]. Such analysis is going to be carried out by explaining the way the consumptions are produced as Habib et al. paper do regarding the garment industry in Bangladesh [3]. The main idea is to map the entire process and characterizes it by a mathematic model that can be easily utilized to optimize the global resources input. The global model consists of single sub-models, which can interact with each other exchanging mass or energy flows. The mathematic research has shown that relevant amounts of energy can be exchanged between kilns, dryers and heating units in order to reduce the primary heating energy demands, which result in an important CO_2 emission reduction [4]. Besides, the thermodynamic state of some wastes is more appropriate to be reused in some other sub-process. It can be seen that recoverable heat from one consumer could be profited into another consumer. Some industrial processes demand a thermal level that can be provided with recovered heat from a consumer with a higher thermal level [5]. The exhausting flow from it is characterized by m and T. This is not a new concept. The real novelty is the methodology to, gradually; apply it to an industrial process where this thermal staggering is driving the production. This characterization has been made possible by the mathematical modeling of the process which is where the innovation of this research lies in.

From this point, they could be stated the different stages to progressively reduce the thermal requirement by means of a physical and mathematical model; it quantifies the energy reduction in each stage, and finally proves that the potential averaged thermal savings are within 40 and 50%.

This book aims to provide a generalized methodology to quantify the savings through a sequence of heat recovering techniques that progressively improves the thermal energy consumption of the factory [6] so as the way to reduce electricity and water global consumptions by means of different ways which include heat utilization. Simulation of the factory from the consumption perspective is a novel way to approach the sanitary-ware industry and its sustainability. Some other fields have been approached in a similar way [7, 8].

This methodology, extendible to any industrial facility where thermal consumption at different thermal levels is present, is particularly applied to the sanitary-ware industry. A proper energy management should consider this technique in any field where thermal power is demanded and wasted [9].

As stated in Chap. 2, they will be considered plaster mold-based factory type versus resin mold type as main developed current technologies [10]. Establishing this basis as a starting point and numerically modeling the consumption of every process to be described in terms of energy and water consumption, they will be calculated as the CO_2 emissions. From these results, the assessment will be constructed, which has to be the main tool to design the most efficient factory in both types so as to quantify the potential energy and water savings [11]. It will be done through a case study as an example to understand in practice how the procedure can be applied to any kind of factory. Some authors have researched strategies for efficiency in energy and resources use for other industrial sectors like Ingarao has done for metal shaping [12]. Ingarao suggests reducing overall CO_2 emissions by tackling the manufacturing sector as a relevant contribution to them. He describes the role of metal shaping processes in reducing the environmental impact across different stages of the metal components' life.

Although resource reduction is a very common practice and it is widely researched [13], the last point (reusing wastes among different processes) is not so extended within the sanitary-ware industry from a scientific point of view. After a comprehensive review, no research has been found about this matter within the sanitary-ware manufacturing, though some applications have been found referring to other fields of manufacturing [14]. Some research has been developed about this matter, they are not focused on a particular sector of activity [15], although all of them are concerned about future scarcity [16]. On the other hand, there is some research regarding a specific industrial sector like Madan et al. for the injection molding industrial sector [17].

This book shows how sanitary-ware manufacturing can be re-engineered to save most of the water and more than almost 1/3 of primary energy.

These relevant values are worth taking into account and study them with deep detail in order to find out the ways to reduce the different energy consumptions involved throughout the global process as some authors have suggested to integrate sustainability indicators and lean manufacturing to assess manufacturing processes [18].

In Sect. 2.2, the two types of factory to be analyzed were defined from a qualitative point of view. In order to assess consumptions and be able to compare each other, it has to be established a quantitative definition. In other words, a size of factory has to be set. A quite standard size of ceramic sanitary-ware factory is that to produce a half million pieces per year, which is consistent with the fact that this is a very common size of tunnel kiln. It can be assumed for the sake of simplification than it is a single product factory. To compute resource requirements, it has to be set the amount of product to be manufactured as well. For the purpose of modeling both factory types, Table 3.1 states the production parameters that this book will consider from this point on.

These parameters represent the starting point to compute requirements in every single process by the appropriated mathematic model.

Figures 3.1 and 3.2 show typical layout for a standard factory in both configurations. It is remarkable the only difference lies in molds making and casting.

3.2 Quantification of Energy and Water Requirements

The starting point to quantify the potential reduction is the assessment of the current consumptions of resources in conventional factories. Consumptions are calculated by means of modeling every process through the entire production line for the two standard technologies and, therefore, calculating the involved resources.

Table 3.1 Production parameters of factory type

Production parameters		
Product=	Toilet seats	
Working days=	232	Days/year
Kiln days=	330	Days/year
Shifts=	1	Shift/day
Hours/shift=	8	Hour/shift
Averaged weigh=	20	kg
Production=	500,000	pieces/year
Global yield=	80.0%	
Casting yield=	95.0%	
Kiln yield=	84.2%	
Electricity price=	0.10	€/kWh
Gas price=	0.03	€/kWh
Water price=	2.00	€/m^3

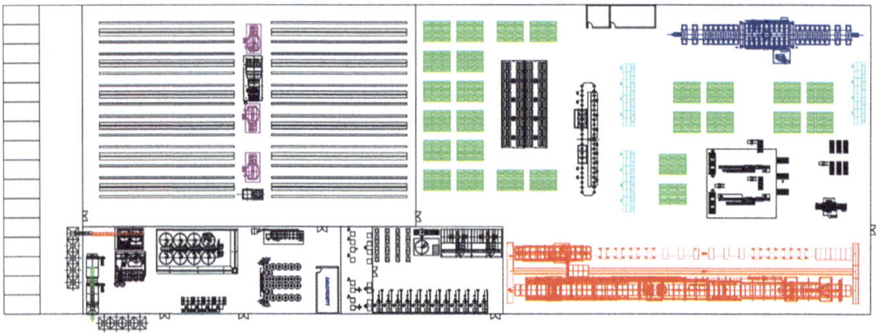

Fig. 3.1 Plaster mold factory type (Courtesy of SACMI GROUP)

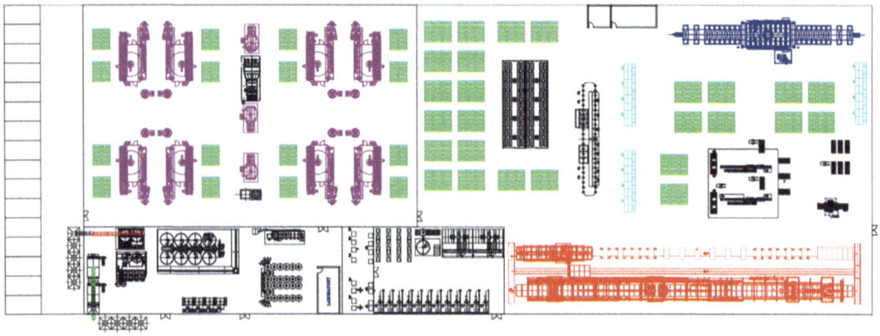

Fig. 3.2 Resin mold factory type (Courtesy of SACMI GROUP)

3.2.1 *Electrical Demand*

Electricity is basically used as:

- Motor power to achieve mechanical actions.
- Lighting throughout the factory.
- Heating for some simple processes.

So, the electrical physical model will be as Fig. 3.1 shows (Fig. 3.3).

3.2.1.1 Electrical Demand for Compressed Air

Regarding pneumatic equipment, it can be considered as an extension of an electrical device. From the perspective of this book, any consumption of compressed air can be translated into an electrical one through a factor that equals the performance rate of the compressor and network. According to Fig. 3.4:

The theoretical specific work for an ideal adiabatic compression is calculated by:

$$\frac{W}{V_1} = P_1 \cdot \left[-1 + r_c^{\frac{\gamma-1}{\gamma}} + \frac{r_c^{\frac{\gamma-1}{\gamma}} - 1}{\gamma - 1} \right] \tag{3.1}$$

where:

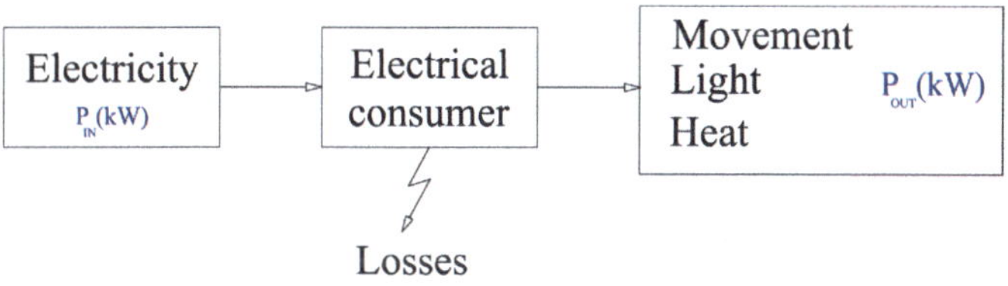

Fig. 3.3 Electrical consumer scheme (*Source* Data compiled by the authors)

Fig. 3.4 Scheme of an ideal adiabatic compression cycle

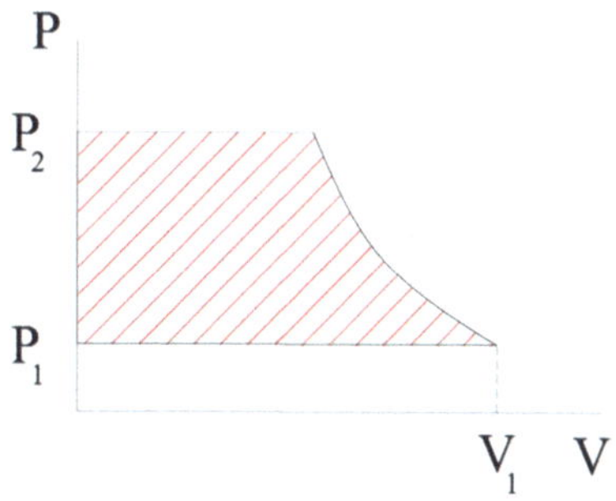

P_1: Inlet pressure at atmospheric conditions

$V1$: Inlet volume at atmospheric conditions

r_c: Compression rate.

In order to obtain the actual electrical consumption, this value has to be divided by the compressor efficiency.

3.2.1.2 Electrical Demand for Lighting

Required lighting is defined by standards on illumination safety levels. The different areas within the factory must reach a minimum level of illuminance and quality of the light, depending on the activity. The model to simulate the electrical demand takes into account the surface to illuminate, so as the illuminance level according to:

$$W_{\text{electrical}} = \frac{S \cdot C_{\text{use}} \cdot E}{\eta} \tag{3.2}$$

where:

$W_{\text{electrical}}$: required electrical power for lighting (kW)

h: efficiency of the light type (lm/kW)

S: surface to illuminate (m^2)

C_u: coefficients of use for the light fixture (dimensionless)

E: illuminance (lx).

3.2.2 Thermal Demand

Basic thermal process is assumed to be a conventional combustion as Fig. 3.2 shows (Fig. 3.5).

This basic process will be combined with other different actions involving other fluids in order to model specific thermal processes along the manufacturing line.

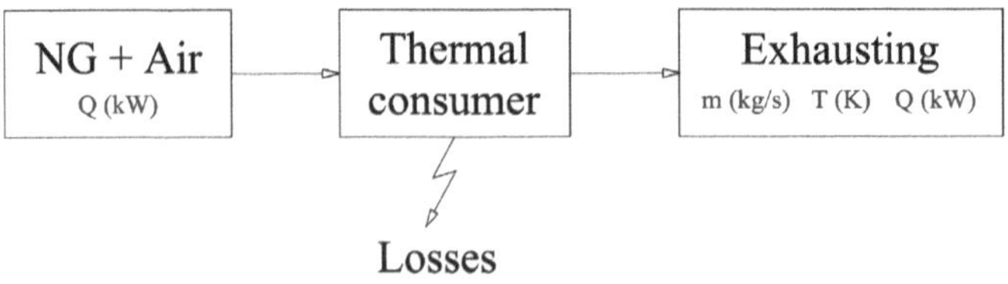

Fig. 3.5 Thermal consumer scheme (*Source* Data compiled by the authors)

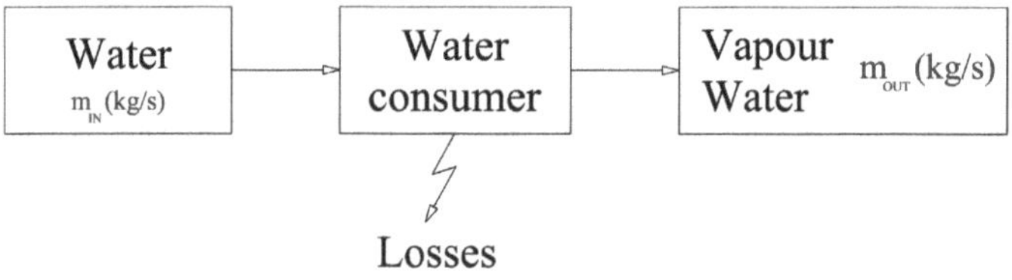

Fig. 3.6 Water consumer scheme (*Source* Data compiled by the authors)

Regarding water, consumption scheme can be considered as per Fig. 3.3 (Fig. 3.6).

Models take into account mass and energy balances, so a system of equations is resolved for each process yielding an output of energy and mass wastes as shown in (3.1) and (3.2):

$$\sum \dot{m}_{\mathrm{IN}} = \sum \dot{m}_{\mathrm{OUT}} \quad (3.3)$$

$$W_{\mathrm{electrical}} + W_{\mathrm{physical}} + W_{\mathrm{fuel}} = W_{\mathrm{products}} + W_{\mathrm{losses}} \quad (3.4)$$

where:

- $\dot{m}_{IN}$, are the mass flows in kg/s of different fluid inlets; air, water, vapor, natural gas, etc.
- $\dot{m}_{OUT}$, are the mass flows in kg/s of different fluid outlets; CO_2, vapor, water, air, etc.
- $W_{\mathrm{electrical}}$, W_{physical}, W_{fuel}, W_{products}, W_{losses}, are the energy flows and losses.

The equations are established considering the thermodynamic states of each flow at both, inlet and outlet, including the losses and inefficiencies. Figure 3.7 shows the methodology of the modeling procedure

- W_{el} is directly defined by the motor power and efficiency.
- W_{ph} refers to physical inputs which may be solid, liquid or gaseous.
- W_{fuel} refers to fuel heating value.
- W_{products} refers to physical outputs streams which may be solid, liquid or gaseous.
- W_{losses} refers to thermal losses along the process.

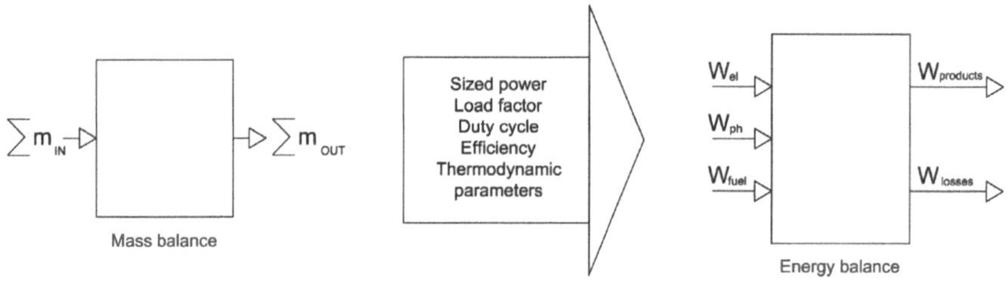

Fig. 3.7 Scheme of the models of processes (*Source* Data compiled by the authors)

3.2.2.1 Calculation of W_{ph}

Normal conditions (1 atm = 101,325 Pa and 273 K) are taken as the reference state for the enthalpies. Thus, the enthalpy of the solids can be taken as:

$$Q = m \cdot Ce \cdot T \tag{3.5}$$

and the heat required to change this temperature:

$$Q = m \cdot Ce \cdot \Delta T \tag{3.6}$$

where:

- m: mass flow (kg/s).
- Ce: Specific heat (kJ/kg/K).
- T: Temperature of the solid at one defined state (K).
- ΔT: Temperature difference at two defined states (K).

For the case of liquids (basically water) or gases, the method is different. The required heat is calculated via enthalpies at the computed physical states. Enthalpies of flows are obtained by the NASA Glenn Research Center database of thermodynamic data JANAF coefficients [19] as:

$$\frac{H(T)}{RT} = -a_1 T^{-2} + a_2 \frac{\ln T}{T} + a_3 + a_4 \frac{T}{2} + a_5 \frac{T^2}{3} + a_6 \frac{T^3}{4} + a_7 \frac{T^4}{5} + \frac{b_1}{T} \tag{3.7}$$

where a_i ($i = 1 \ldots 7$) and b_1 are tabulated for each component.

Thus, the required heat to from one to another state is calculated by the difference of enthalpies.

3.2.2.2 Calculation of W_{losses}

A simple approach of heat transfer through the walls and roof of the drier provides the estimations of thermal losses during the process.

$$Q_{\text{losses}} = U \cdot S_{\text{enclosure}} \cdot \Delta T \tag{3.8}$$

where:

- U: Global heat transfer coefficient (kJ/m^2/K).
- $S_{\text{enclosure}}$: Total surface of walls and roof (m^2).
- ΔT: Temperature difference between the chamber and hall environment (K).

3.2.3 Psychrometric Calculations

Evaporation is a constant within all thermal processes along the factory. Thus, the model to carry out mass and energy balances linked to this change of state is described as follows

The saturation pressure of the vapor at a given temperature is calculated by Sonntag's formula [20]:

$$p_s = \exp\left(\sum_{i=1}^{4} a_i \cdot T^{i-2} + a_7 \ln(T)\right), \tag{3.9}$$

where:

$a1=$	−6.09693850E+03
$a2=$	2.12409642E+01
$a3=$	−2.71119300E-02
$a4=$	1.67395200E-05
$a7=$	2.43350200E+00
p_s	(Pa)
T	(K)

with a standard uncertainty <0.005%rh.

Humidity ratio of moist air can be expressed as:

$$x = \frac{m_v}{m_a} \cdot \frac{kg_{\text{vapor}}}{kg_{\text{air}}}. \tag{3.10}$$

where:

x = Water content (kg/kg).

m_v = Vapor mass (kg).

m_a = Air mass (kg).

Correspondent masses of vapor and air are expressed as:

$$m_v = \frac{M_w \cdot p_v \cdot V}{R_0 \cdot T}; \tag{3.11}$$

$$m_a = \frac{M_a \cdot p_a \cdot V}{R_0 \cdot T}; \tag{3.12}$$

where:

M_w = molecular mass of vapor (kg/kmol).

p_v = partial pressure of vapor (Pa).

V = volume of the vapor-air mixture (m^3).

R_0 = ideal gas constant (kJ/K/kmol).

T = temperature of the mixture (K).

M_a = molecular mass of air (kg/kmol).

p_a = partial pressure of air (Pa).

Hence:

$$x = \frac{m_v}{m_a} = \frac{M_v \cdot p_v}{M_a \cdot p_a}. \tag{3.13}$$

Relative humidity can be expressed as:

$$\phi = \frac{p_v}{p_s} \tag{3.14}$$

and total pressure of the mix:

$$p = p_v + p_a. \tag{3.15}$$

Thus, combining (3.5)–(3.7):

$$x = \frac{M_v}{M_a} \cdot \frac{p_s}{\frac{p}{\phi} - p_s} \frac{kg_{\text{vapor}}}{kg_{\text{air}}}. \tag{3.16}$$

At saturation pressure,

$$\phi = 1$$

and

$$x_s = \frac{M_v}{M_a} \cdot \frac{p_s}{p - p_s} \frac{kg_{\text{vapor}}}{kg_{\text{air}}}. \tag{3.17}$$

Then, actual water content is provided by relative humidity of the environment conditions with:

$$x = x_s \cdot \phi \cdot \frac{\text{kg}_{\text{vapor}}}{\text{kg}_{\text{air}}}. \tag{3.18}$$

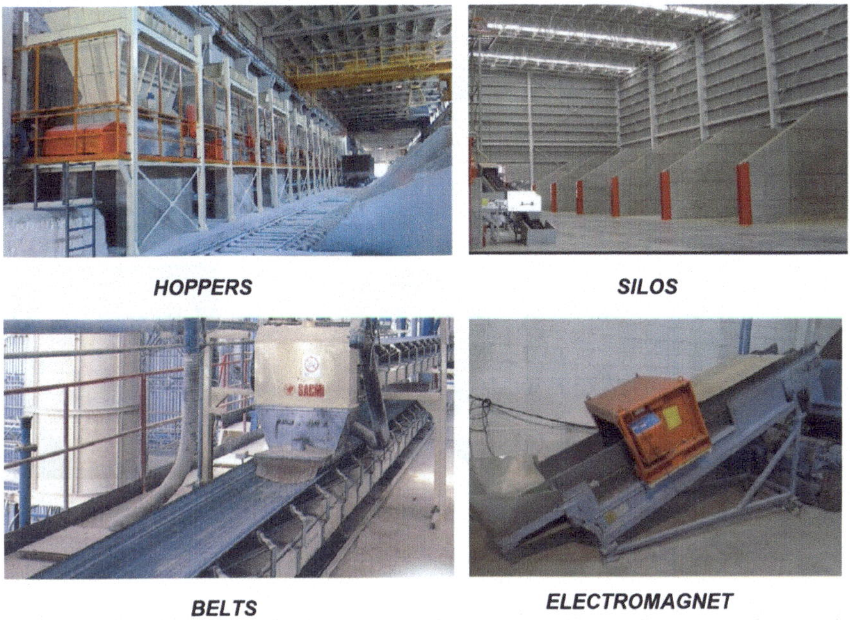

Fig. 3.8 Equipment for raw materials (Courtesy of SACMI GROUP)

3.3 Requirements in Every Section of the Factory

Requirements for the different processes in every section according to the structure of the manufacturing line stated in Sect. 2.2 and the ceramic sanitary-ware manufacturing general process defined in Sect. 1.5 are briefly explained as follows:

3.3.1 Raw Materials

- Belts for conveying raw materials between silos and mills, blungers, hoppers, sieves, etc. Electric motors are required to power them.
- Hoppers for dosing and storing materials. Electrical or pneumatic devices are required to power agitation automatic valves, etc.
- Magnets to separate ferric particles. They may be electrically supplied (Fig. 3.8).

3.3.2 Slip Preparation

- Mills to get hard materials (silica, quartz, feldspar) micronized. Powerful electric motors are required to run these heavy rolls. They can be conceived either by

Fig. 3.9 Equipment for slip preparation (Courtesy of SACMI GROUP)

humid or dry via. Usually, they start dosing stage by adding part of the water corresponding formula.

- Belts for conveying soft raw materials into blungers. Electric motors required to power them.
- Blungers to finish dosing stage, mixing components and homogenize slip. Powerful electric motors are required to mix slip with high density and viscosity at a high speed.
- Sievers to control particles size and eliminate possible contamination. They are usually equipped with electric vibrators.
- Pumps to transfer flows between mills, blungers or tanks throughout the factory. When slip is transferred to short distances not much pressure is required and pneumatic pumps are used. On the contrary, when higher pressure is required, electric peristaltic pumps are commonly used.
- Slip tanks to mature and age slip before being supplied to casting workshop. Slip must be under agitation in order to avoid decantation and maintain correct homogenization. Thus, electrical motors are required. They are not so powerful as blunger ones because speed has to be moderate (Fig. 3.9).

3.3.3 Glaze Preparation

The process to prepare the glaze is quite likely slip, differing only in formula and quantities. Equipment is notably smaller and less powerful, but similar conceptually.

Fig. 3.10 Equipment for glaze preparation (Courtesy of SACMI GROUP)

- Mills
- Belts
- Blungers
- Sieves
- Pumps
- Glaze tanks (Fig. 3.10).

3.3.4 Modeling and Molds

Plaster molds factory

- Hopper to collect plaster. Electrical or pneumatic devices are required to power agitation and automatic valves.
- Screw conveyor to supply plaster powder from hopper to mixer. Electric motor is required to power it.
- Plaster mixer to arrange the mixture of water and powder powered by electric motor.
- Dryers to dry the finished mold after casted. Electric fans to provide air movement and ventilations are required so as heating required to increase the psychrometric capacity of thc air.

Fig. 3.11 Equipment for modeling and molds (Courtesy of SACMI GROUP)

- Cranes to move molds and matrixes which may reach hundreds of kilograms.
- Pumps for water might be required.

Resin molds factory

This equipment is required to manufacture resin molds but it is required as well to manufacture matrixes which are used to cast plaster molds, so it is present also within plaster molds factory.

- Mixer to arrange the mixture of resin components powered by electric motor.
- Cranes to move molds and matrixes, which may reach hundreds of kilograms.
- AC Unit to provide the right environmental temperature preventing ignition of some components so as to regulate proper curation of resins hardening.
- Electric pumps to provide water.
- Ventilation fans to provide air renovation powered with electricity.
- Stoves to heat up some components of resins. Usually powered by electric resistors (Fig. 3.11).

3.3.5 *Casting*

Plaster molds factory

- In Casting benches, casting is carried out. Essentially, they consist of a bench with molds on top, but some accessory requirements are needed to achieve the process.
  - Local lighting
  - Compressed air to
  - Compressed air to blow mold after demolding
  - Compressed air to keep pressure during casting procedure
  - Compressed air for demolding manipulator requirements

  - Low voltage electric supply for demolding manipulator
  - Electrical fans to blow molds during drying process of plaster
  - Electrical motors of demolding manipulator so as lifting devices for mold cover lids.

- Ponds to recover overflows of slip and slip returns. They are equipped with electric pumps.
- Slip tanks to supply the molds for casting. Electric motors required to keep agitation against decantation.
- Electrical and pneumatic pumps to transfer slip and water are required.

Resin molds factory

- Pressure casting machines require:

  - Electricity for closing motorized devices
  - Electricity for hydraulic pressure system
  - Compressed air to blow molds during and after casting cycle
  - Desalted water to flush the mold after casting cycle.

- Boilers are required to warm up the water to flush and clean the molds after every cast cycle.
- Ponds to recover overflows of slip and slip returns. They are equipped with electric pumps.
- Slip tanks to supply the molds for casting. Electric motors required to keep agitation against decantation.
- Electrical and pneumatic pumps to transfer slip and water are required (Fig. 3.12).

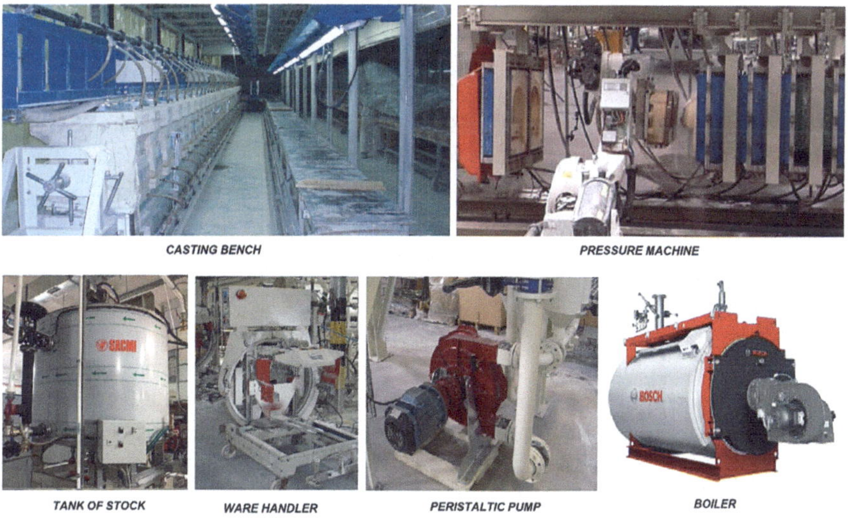

Fig. 3.12 Equipment for casting (Courtesy of SACMI GROUP)

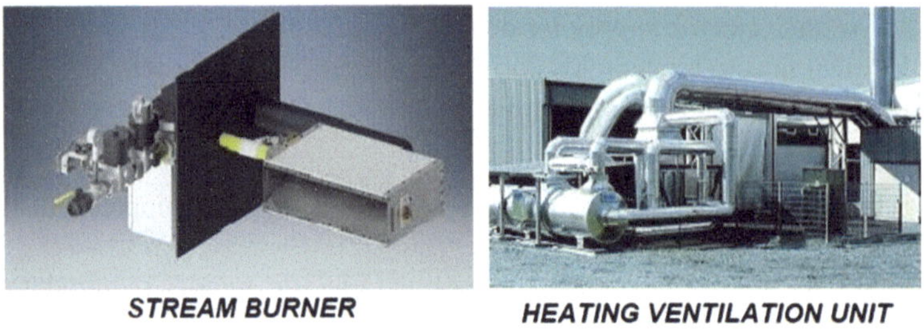

Fig. 3.13 Equipment for heating and ventilation (Courtesy of SACMI GROUP)

3.3.6 Heating and Ventilation

- Heating and ventilation units (HVU) are required to keep the appropriate hygrothermal conditions within casting hall, where temperature and humidity conditions are crucial for the quality of the manufacturing. Equipment parts for these units include:
  - Powerful electrical fans to blow air into the hall
  - Gas line burners to heat up inlet air either from hall or external.
  - Water spray nozzles are required when blown air is to dry according to set up point of humidity.
- Exhausting fans remove saturated air in order to allow evaporation process within hall environment (Fig. 3.13).

3.3.7 Drying

- Dryers operation seems quite like HVU's. Air is blown and recirculated within the dryer chamber. Temperature is required to increase the dew point of the air and exhausting fans are in charge of removing saturated air (Fig. 3.14).

3.3.8 Glazing

- A carrousel is a complex equipment where many electrical motors are required for moving chains, conveyors, fans, etc. Compressed air and electricity are required as well to arrange automation of devices by means of cylinders, valves, etc. On the other hand, compressed air is also massively consumed at this point due to the need of blowing ware after grinding and polishing.
- Booths and others. They act similarly to a carrousel in a static way where ware is not under movement. Only rotation motors and ventilation fans are required as well as compressed air for automation and ware blowing (Fig. 3.15).

WARE DRYER

Fig. 3.14 Equipment for drying (*Source* Data compiled by the authors)

GLAZING BOOT **GLAZING CARROUSEL**

Fig. 3.15 Equipment for glazing (Courtesy of SACMI GROUP)

3.3.9 Kilns

- Tunnel kiln is the most demanding equipment from thermal point of view. About half of the total consumption is devoted to it in order to rise ware temperature over 1,000 °C. On the other hand, electricity is very present to move; fans for air impelling and exhausting, carts along the tunnel chamber valves for automation, etc. Compressed air is required for automation.
- Shuttle kiln operates similarly to the tunnel. The main differences lie on the duty cycle and capacity. Shuttle is dedicated to fire only reparable losses under lower temperature demand. For an efficient tunnel kiln, firing losses have to be few. Thus, shuttle kiln is quite less demanding of thermal energy.
- Repairing booths prior to second fire through the shuttle kiln, faulty ware has to be adjusted with electrical or pneumatic hand tools. Ventilation with electrical fans is required (Fig. 3.16).

Fig. 3.16 Equipment for kilns (*Source* Data compiled by the authors)

3.3.10 Sorting

- Test lines require electricity to move conveyors, automation, blowers, etc. Compressed air is demanded to act all automation equipment; cylinders, valves, test stations, etc.
- Wrapping kiln for packaging is heated up by gas to shrink plastic film around the packaged ware. Electricity for conveyors and motors as well automation is also required.
- Packing lifting tables are supplied with both, electricity and compressed air for motors and automation. Many times, a handler acted by compressed air is present at this stage (Fig. 3.17).

Fig. 3.17 Equipment for sorting (Courtesy of SACMI GROUP)

3.3.11 Offices

- Desk station with computer, printer, and other similar electrical equipment is required at the offices.
- Central server unit with electrical ventilation and conditioned air is required.
- Air conditioning and/or heating units may be used.
- Others; water source, fans, lighting, etc.

3.3.12 Ware-House

- Ware-house office with some desk station as offices, air conditioning and heating, etc.
- Battery chargers for electrical lift trucks are required.

3.3.13 Facilities

- Compressors to supply compressed air through the whole factory to act all the automation around so as provide blowing of ware in different processes.
- Boilers for water warming. Warm water is required either for production purposes (slip warming, resin flushing, quality control, etc.) or sanitary purposes (hand-washing, showers kitchen or canteen, etc.).
- Maintenance area usually disposes a work shop with varied machinery to mechanize and repair equipment. Many electrical or pneumatic equipment are used in this section.
- Reverse osmosis is quite common to desalt water in order to produce slip or flushing resin molds. A great amount of electrical power is required for this process by means of relatively high pressure pumps.
- Losses hopper and crusher require electrical power to move motors and conveyors, to automate the equipment, etc. Compressed air is as well required.
- Pumps room for firefighting and production purposes require electrical power which sometimes it is relatively high due to sprinklers or fire hoses with high discharge capacity.
- Wastewater treatment plant demands electrical power for pumps, motors valves, dampers, flaps, scrapers, etc.
- External lighting.
- Sockets.

3.4 Existing and Potential Links of Resource and Wastes

Industrial experience within the sector states some conclusions which this book intends to demonstrate:

- A great deal of energy is involved
- Thermal energy is about four times the electrical energy
- Electricity consumption is slightly greater in plaster molds manufacturing
- Water consumption is eight times greater in resin molds manufacturing.

The methodology for the progressive application of reengineering is as follows:

- Consumption of standard factories is quantified where no optimizing measures are not considered. This is set as the standard starting point to produce reduction of consumptions.
- Once stated the resources demanded by conventional factories, the following action is to identify the potential reductions. The main clues to find them is to go through the equations systems trying to eliminate as many inputs from the processes as possible. This can be achieved in different ways:
  - Improve the efficiency of the machines by general electrical, thermal and water-saving measures, which are applied as they are usually considered in any industry to achieve any resource performance improvement.
  - Try to set renewable energies.
  - Try to find alternative technologies of every single process with less demand on resources.
- At this point, it is applied the specific set of measures likely appropriated for this industrial sector as:
  - Residual heat from kilns is reconducted towards heat consumers with lower thermal level as dryers, heating units or boilers.
  - Pre-heating combustion air of kilns is as well considered, which reduces their NG consumption [21]. On the other hand, less air for combustion is needed which derives a better efficiency [22].
  - Application of thermal new processes fueled with recovered heat to reduce electrical and water consumptions.
  - Since a conventional factory is run in shifts of 8 h, most of the continuous residual heat from kilns is lost. In order to match heat recovery with thermal requirement and flat the thermal demand curve, the cycles of consumers are extended beyond the standard shift through the whole day long (24 h).
  - Although consumption is improved by flattening the thermal demand curve, it can be further improved by storing heat from periods of heat excess to be used during periods of heat defect.

The mathematic model seen as a tool of optimization is an important scientific advance to design new more efficient factories. It has shown that heat waste of the kiln can be reused to:

- Preheat the air to supply its combustion feeding resulting in less gas consumption.
- Supply dryers.
- Supply heating of casting hall.
- Supply boilers of casting.
- Heat a multi-effect distiller to desalinate water for molds' washing.
- Supply an absorption machine to condense vapor from dryers and casting exhausting flows.

On the other hand, cogeneration or renewable power plants can be coupled to reduce both, electrical and thermal requirements from primary energy so as to provide residual power for lower temperature consumers of heat.

References

1. Maheshwari A, Prasad V, Gudi RD, Biswas P (2019) Systems engineering based advanced optimization for sustainable water management in refineries. J Clean Prod 224:661–676. https://doi.org/10.1016/j.jclepro.2019.03.164
2. Sathitbun-anan S, Fungtammasan B, Barz M, Sajjakulnukit B, Pathumsawad S (2014) Energy efficiency and greenhouse gas emission reduction potentials in sugar production processes in Thailand. Energy Sustain Dev 23:266–274. https://doi.org/10.1016/j.esd.2014.09.010
3. Habib MA, Hasanuzzaman M, Hosenuzzaman M, Salman A, Mehadi MR (2016) Energy consumption, energy saving and emission reduction of a garment industrial building in Bangladesh. Energy 112:91–100. https://doi.org/10.1016/j.energy.2016.06.062
4. Colmenar-Santos A, Zarzuelo-Puch G, Borge-Diez D, García-Diéguez C (2016) Thermodynamic and exergoeconomic analysis of energy recovery system of biogas from a wastewater treatment plant and use in a stirling engine. Renew Energy 88:171–184. https://doi.org/10.1016/j.renene.2015.11.001
5. Delpech B, Axcell BP, Jouhara H A review on waste heat recovery from exhaust in the ceramics industry. In: E3S Web Conference, January 2017, vol 22, p 00034. https://doi.org/10.1051/e3sconf/20172200034
6. Goldemberg J (1994) An energy manual for practicing engineers. Energy Sustain Dev 1(3):59. https://doi.org/10.1016/S0973-0826(08)60056-3
7. Seferlis P, Varbanov PS, Klemeš JJ (2015) Applied thermal engineering solutions through process integration, modelling and optimisation. Appl Therm Eng 89:1001–1005. https://doi.org/10.1016/j.applthermaleng.2015.07.060
8. Wang X, El-Farra NH, Palazoglu A (2017) Optimal scheduling of demand responsive industrial production with hybrid renewable energy systems. Renew Energy 100:53–64. https://doi.org/10.1016/j.renene.2016.05.051
9. Johansson MT (2015) Improved energy efficiency within the Swedish steel industry—the importance of energy management and networking. Energy Effic 8(4):713–744. https://doi.org/10.1007/s12053-014-9317-z
10. Reference document on best available techniques in the ceramic manufacturing industry. European Commission, August 2007 (2007). http://eippcb.jrc.ec.europa.eu/reference/BREF/cer_bref_0807.pdf. Accessed 02 Dec 2016
11. Communication Environmental Website|GREEN STORY|TOTO (2016). http://www.toto.co.jp/greenchallenge/technology/story/14/en/. Accessed 02 Dec 2016
12. Ingarao G (2017) Manufacturing strategies for efficiency in energy and resources use: the role of metal shaping processes. J Clean Prod 142:2872–2886

13. Roldán-Blay C, Escrivá-Escrivá G, Roldán-Porta C, Álvarez-Bel C (2017) An optimisation algorithm for distributed energy resources management in micro-scale energy hubs. Energy 132:126–135. https://doi.org/10.1016/j.energy.2017.05.038
14. Galati M, Iuliano L, Salmi A, Atzeni E (2017) Modelling energy source and powder properties for the development of a thermal FE model of the EBM additive manufacturing process. Addit Manuf 14:49–59
15. Effenberger F, Hilbert A (2016) Towards an energy information system architecture description for industrial manufacturers: decomposition & allocation view. Energy 112:599–605. https://doi.org/10.1016/j.energy.2016.06.106
16. Grubbström RW (2015) On the true value of resource consumption when using energy in industrial and other processes. Int J Prod Econ 170, Part B:377–384. https://doi.org/10.1016/j.ijpe.2015.07.011
17. Madan J, Mani M, Lee JH, Lyons KW Energy performance evaluation and improvement of unit-manufacturing processes: injection molding case study. J Clean Prod 105:157–170 (2014)
18. Helleno AL, de Moraes AJI, Simon AT (2016) Integrating sustainability indicators and lean manufacturing to assess manufacturing processes: application case studies in Brazilian industry. J Clean Prod 1–12
19. McBride BJ, Zehe MJ, Gordon S (2002) NASA glenn coefficients for calculating thermodynamic properties of individual species, p 297
20. Various (2014) Guide to meteorological instruments and methods of observation, 8th edn. World Meteorological Organization
21. Mezquita A, Boix J, Monfort E, Mallol G (2014) Energy saving in ceramic tile kilns: cooling gas heat recovery. Appl Therm Eng 65(1):102–110. https://doi.org/10.1016/j.applthermaleng.2014.01.002
22. Shiralkar KY, Kancharla SK, Shah NG, Mahajani SM (2014) Energy improvements in jaggery making process. Energy Sustain Dev 18:36–48. https://doi.org/10.1016/j.esd.2013.11.001

Chapter 4
Production Line: Process and Energy Modeling

According to the general process as per Chap. 2, the factory has been modeled as a sequence of sections agreeing with the specified processes. Depending on the process, they will be stated the general requirements of energy and water described in the following sections.

4.1 Raw Materials Section

Common components in sanitary-ware formulation are: clays and kaolin as plastic materials and quartz, feldspar and chamottes as hard materials. These materials are usually provided as a mixture of different components as shown in Tables 4.1, 4.2, 4.3 and 4.4.

In order to calculate energy requirements to deal with raw materials, it is important to assess some fundamental variables related to them such as density, which is calculated as per composition as shown in Table 4.5 with averaged densities.

From operation factory parameters seen in Table 3.1, they can be calculated the amounts of raw materials needed to produce the ware and thus the size of equipment in terms of power.

In order to produce the final output of 500,000 pieces in 1 year, the weight of raw materials should be:

$$P \cdot w \cdot \frac{1}{\eta} = m_{\text{solids}} \tag{4.1}$$

where:

P: Production of ware (pieces/day)

W: weigh of every piece (kg/piece)

η: production global yield [0.8].

C. Cuviella-Suárez et al., *Water and Energy Use in Sanitary-ware Manufacturing*, Green Energy and Technology, https://doi.org/10.1007/978-3-030-72491-7_4

Table 4.1 Usual composition of ball clays [1]

	Density	Ball clays			
	kg/m^3	Hycast-VC	Hycast-Zodiact	Sanblen 90	Sanblend GC
SiO_2	2,650	52.4%	54.0%	60.6%	63.3%
Al_2O_3	3,950	29.7%	28.5%	24.1%	24.0%
Fe_2O_3	5,240	1.2%	1.2%	0.9%	0.9%
TiO_2	4,230	1.0%	1.0%	1.2%	1.5%
CaO	3,340	0.2%	0.2%	0.1%	0.1%
MgO	3,580	0.3%	0.3%	0.3%	0.3%
K_2O	2,350	1.8%	1.8%	1.8%	1.8%
Na_2O	2,270	0.2%	0.2%	0.3%	0.3%
(*)IL	1,000	11.4%	11.1%	9.2%	7.4%
C	2,250	1.9%	1.9%	1.6%	0.5%
Total		100%	100%	100%	100%
Density	kg/m^3	2,885	2,874	2,843	2,881

(*) IL: "ignition losses" which refers to organic volatile components that are burnt in the firing process

Table 4.2 Usual composition of kaolin [1]

	Density	Kaolin			
	kg/m^3	Remblend	NSC	Kernick	31
SiO_2	2,650	48.0%	48.0%	48.0%	48.8%
Al_2O_3	3,950	36.5%	36.5%	36.5%	35.4%
Fe_2O_3	5,240	1.0%	0.9%	1.0%	0.8%
TiO_2	4,230	0.1%	0.1%	0.1%	0.1%
CaO	3,340	0.1%	0.1%	0.1%	0.1%
MgO	3,580	0.3%	0.3%	0.3%	0.2%
K_2O	2,350	2.0%	2.0%	2.3%	2.8%
Na_2O	2,270	0.1%	0.1%	0.2%	0.1%
IL	1,000	12.0%	11.9%	11.6%	11.6%
C	2,250	0.0%	0.0%	0.0%	0.0%
Total		100%	100%	100%	100%
Density	kg/m^3	2,951	2,950	2,956	2,935

From Table 4.5, 53.5% of solids are hard materials to be milled before mixing with plastics in the blunger. So, the size of the mill has to be over:

$$m_{\text{solids}} \cdot 53.5\% = m_{\text{hardsolids}}. \tag{4.2}$$

Table 4.3 Usual composition of feldspars [1]

	Density	Feldspar	
	kg/m^3	Feldspar Na	Feldspar K
SiO_2	2,650	68.7%	64.8%
Al_2O_3	3,950	19.5%	18.3%
Fe_2O_3	5,240		
TiO_2	4,230		
CaO	3,340		
MgO	3,580		
K_2O	2,350		16.9%
Na_2O	2,270	11.8%	
IL	1,000		
C	2,250		
Total		100.0%	100.0%
Density	kg/m^3	2,859	2,837

Table 4.4 Usual composition of chamottes [1]

	Density	Chamottes	
	kg/m^3	Low expansion	High expansion
SiO_2	2,650	56.9%	72.1%
Al_2O_3	3,950	38.5%	24.5%
Fe_2O_3	5,240	1.5%	1.2%
TiO_2	4,230	1.6%	1.2%
CaO	3,340	0.3%	0.2%
MgO	3,580	0.3%	0.2%
K_2O	2,350	0.8%	0.5%
Na_2O	2,270	0.1%	0.1%
IL	1,000		
C	2,250		
Total		100%	100%
Density	kg/m^3	3,217	3,020

Before mixing with the plastic materials, the hard ones must to be micronized by means of milling process which constitutes the major electrical power of this section.

By means of Bond's law it can be calculated the total energy to reduce the diameter of the particle depending on its hardness.

$$E = 10 \cdot W_i \cdot \left(\frac{1}{\sqrt{d_{\text{outlet}}}} - \frac{1}{\sqrt{d_{\text{inlet}}}} \right) \cdot \left[\frac{\text{kWh}}{t} \right] \tag{4.3}$$

where:

Table 4.5 Formula and density of slip raw materials [1]

Slip formula			rho kg/m^3
Plastic materials	Ball clay	26.0%	2,871
	Kaolin	27.5%	2,948
	sub-Total=	53.5%	2,910
Hard materials	Quartz	9.5%	2,650
	Feldspar	29.5%	2,848
	Chamotte	7.5%	3,118
	sub-Total=	46.5%	2,851
	TOTAL=	100%	2,883

- W_i: Work index (14 for quartz, 12 for feldspar)
- $d_{\text{outlet}}, d_{\text{inlet}}$: outlet and inlet particle diameters, respectively (2 and 15 mm).

The milling process is based on the relative movement of grinding pebbles inside falling down in a sort of cascade making grinding pebbles crashing each other and the particles of product.

The torque exerted by the mill can be calculated from the composition of gravity and friction forces as seen in Fig. 4.1:

$$T = r_g \cdot mg \cdot sen\alpha + r \cdot \mu \cdot mg \cos\alpha \tag{4.4}$$

and thus, the power results:

$$N = \omega \cdot T = \left(\omega \cdot r_g \cdot mg \cdot sen\alpha + r \cdot \mu \cdot mg \cos\alpha\right). \tag{4.5}$$

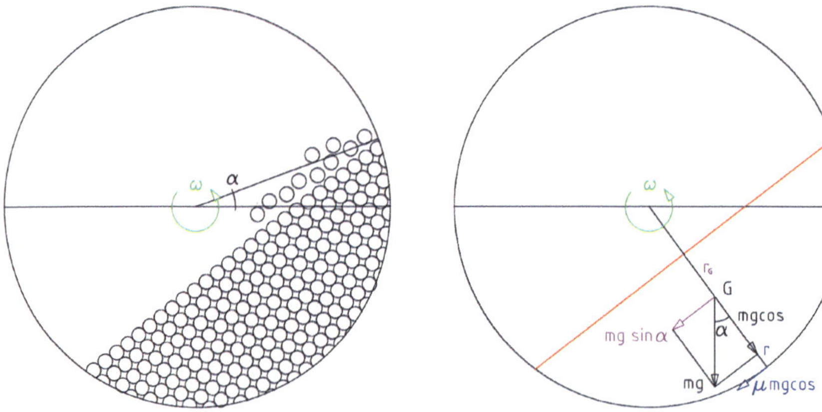

Fig. 4.1 Forces acting on the mill rolling

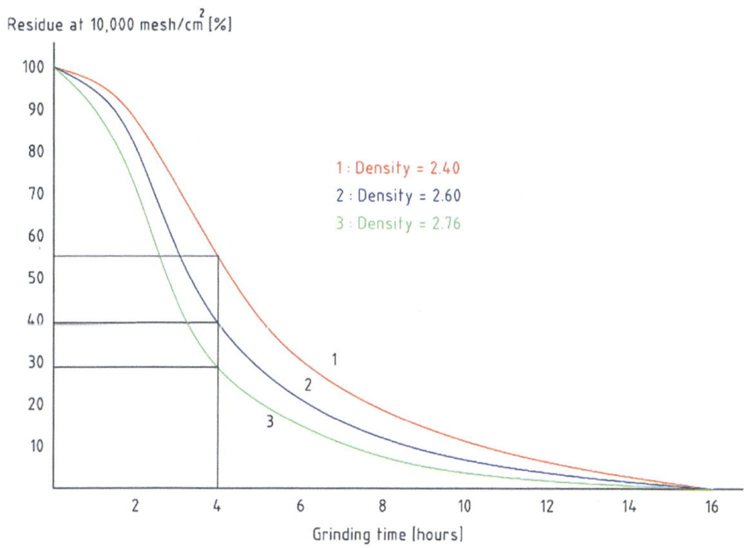

Fig. 4.2 Milling times dependency on the density of grinding media [1]

Milling time depends on different factors as density and quantity of grinding media.

In Fig. 4.2, three different densities are compared to assess the differences of milling time based on 24 h for zero residue at 10,000 mesh/cm^2. This sieve size has a span of 62 μ.

In the graphic, it can be seen how different is the particle size distribution after 4 h of operation depending on the media density. Milling common times are in the range of 6–8 h. However, this time depends strongly on the original size. The required range of power can be estimated with the required energy to mill the materials as:

$$N = \frac{E}{t} \tag{4.6}$$

where:

N: Mechanical power to mill (kW)

E: Mechanical energy to mill as per (4.3) (kJ)

t: milling cycle (s).

In Fig. 4.3, power required and milling time are plotted versus quantity of grinding media load.

In practice, pebbles level recommended is 50–55% of the mill volume.

Obviously, the slip fills the gaps between grinding balls, so there are three different zones in the cross-section of the mill as Fig. 4.4 shows:

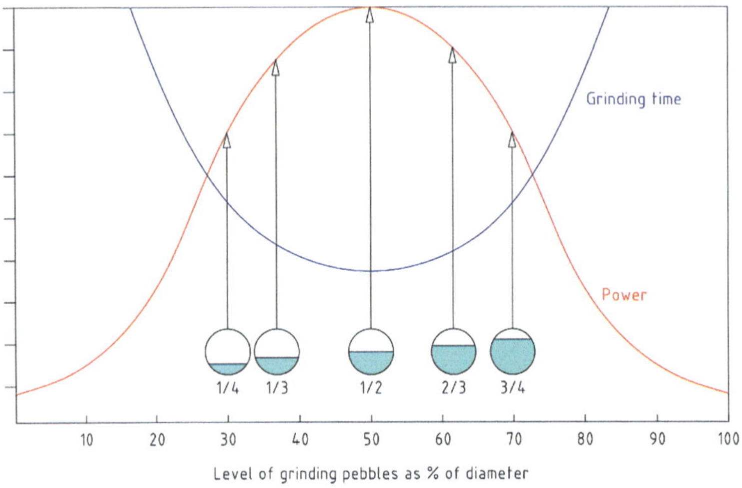

Fig. 4.3 Milling times dependency on the quantity of grinding media [1]

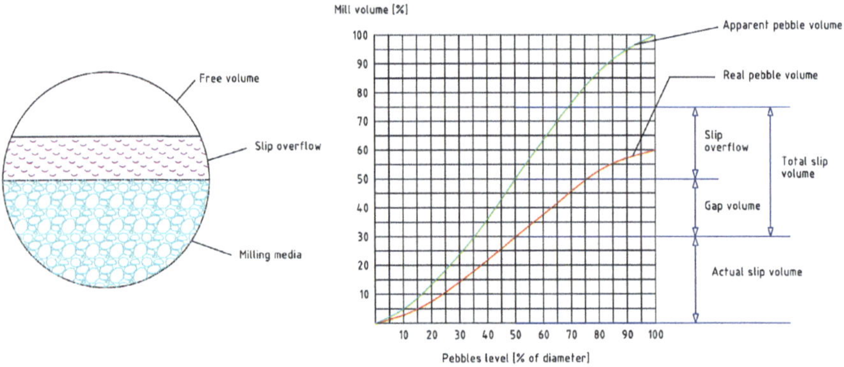

Fig. 4.4 Relationship between different volumes, real volume and level of grinding pebbles [1]

As shown, pebbles level of 50% and total mill volume of 75% implies 45% of the mill volume filled with slip.

The relationship between the density of the slip and components is defined by the following system of equations:

$$v_{\text{slip}} = \frac{m_{\text{slip}}}{\rho_{\text{slip}}} \tag{4.7}$$

$$v_{\text{water}} = \frac{m_{\text{water}}}{\rho_{\text{water}}} \tag{4.8}$$

$$v_{\text{solids}} = \frac{m_{\text{solids}}}{\rho_{\text{solids}}} \tag{4.9}$$

Table 4.6 Composition of hard materials slip depending on the density

Slip of hard materials (water/solids ratio)	Density	Mass	Volume
	kg/m^3	kg	m^3
Slip of hard materials	1,800	36,599	20.33
Water	1,000	11,546	11.55
Solids	2,851	25,054	8.79

$$\upsilon_{\text{slip}} = \upsilon_{\text{water}} + \upsilon_{\text{solids}} \tag{4.10}$$

$$m_{\text{slip}} = m_{\text{water}} + m_{\text{solids}}. \tag{4.11}$$

With υ for volume, m for mass and ρ for density.

The quantity of solid materials within the mill percentage is defined by the selected density of the slip. A common composition of hard materials slip is shown in Table 4.6:

To calculate the total volume of milling facilities, the slip volume is divided by 0.45 as per Fig. 4.4:

$$\frac{m^3_{\text{slip}}}{0.45} = m^3_{\text{workingvolume}} \tag{4.12}$$

Table 4.7 shows common specifications for industrial mills.

Usually, milling facilities included more than one mill to reduce milling time and the mill size. Besides, there is a spare in the event of one mill fails.

Other electrical equipment within this section is run by smaller motors in the range of some kW's for belts, hoppers, etc. Fig. 4.5 shows the general scheme for this section.

4.2 Slip Preparation

The most relevant consumption within this section lies in the blungers.

The key values to define the slip in order to meet the rheological and thixotropic specifications for production are density and viscosity.

Assuming a value of 1,795 kg/m^3 for the density of the barbotine as a very common value in the sanitary industry, the ratio water/solids results as per Table 4.8 (Fig. 4.6).

Common sizes of current industrial blungers are shown in Table 4.9.

Blunger cycle time ranges 5–6 h. Usually, it is preferred to dispose more than one unit to optimize batches of production and have spare unit in the event of any failure.

Up to this point, it has been considered only fresh slip for 1-day production. Actually, the amount of slip to be moved along the factory is greater because of

Table 4.7 Common mill specifications [2]

		MILLS									
		MTD-002	MTD-005	MTD-010	MTD-020	MTD-050	MTD-120	MTD-160	MTD-340	MTD-380	MTD-450
Diameter	mm	730	975	1,112	1,420	1,820	2,390	2,624	3,180	3,180	3,180
Length	mm	600	804	1,050	1,415	1,980	2,763	3,150	4,300	4,880	5,680
Power	kW	0.75	2.2	5.5	11	22	55	75	110	132	160
Speed	r.p.m.	47.5	39	34	24	18	17	16	12.5	12.5	12.5
Working volume	m^3	0.14	0.39	0.80	1.80	4.44	10.66	14.9	30.69	35.2	41.1
Weigh of cylinder	kg	185	300	790	1,010	1,915	5,490	6,630	12,000	14,700	16,800
Weigh of liner	kg	232	439	700	1,180	2,120	5,080	6,430	10,230	11,090	13,000
Weigh of alumina balls	kg	160	450	950	2,100	5,200	11,420	16,015	19,800	22,500	26,300

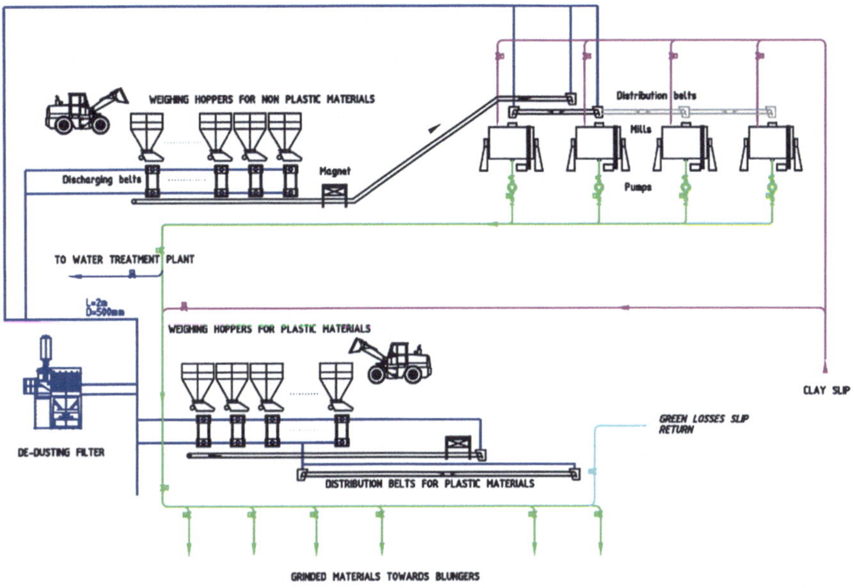

Fig. 4.5 Raw materials general scheme

Table 4.8 Composition of the barbotine and total daily volumes

Barbotine (water/solids ratio)	Density	Mass	Volume
	kg/m^3	kg	m^3
Barbotine	1,795	79,454	44.26
Water	1,000	25,574	25,57
Solids	2,883	53,879	18.69

the returns. On the other hand, fresh slip has to be matured up to 3 days at least before it is sent to casting area. Thus, some stock of slip has to be allowed for. During casting cycle, the wall that conforms the product is stuck to the mold until the desired thickness is reached. At this point, the mold is opened and all the liquid slip remaining inside is released a return of barbotine to be used for a new cycle. This is called the slip return. Usually, it is approximated 2.5–3 times fresh slip. Summarizing the total stock volume has to be at least:

$$m^3_{\text{freshslip}} + 3 \cdot m^3_{\text{maturingslip}} + 3 \cdot m^3_{\text{returnslip}} = 7m^3_{\text{freshslip}}. \tag{4.13}$$

A quite common volume for this kind of tanks is about 50 m^3. Yet some more space is allowed for the handling of slip flows between the different tanks, it can be estimated that seven tanks are needed. Slip cannot remain still in order to avoid decantation. Thus, there is a need of moving it by rotating blades as shown in Fig. 4.7.

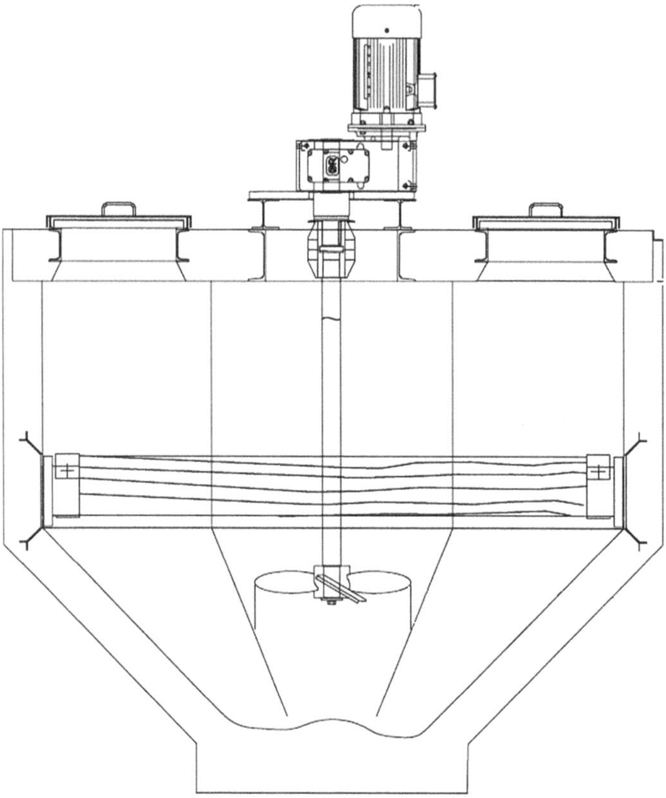

Fig. 4.6 Turbo blunger scheme [2] (*Source* Data compiled by the authors)

Table 4.9 Turbo blunger specifications [2]

		TDT-007	TDT-010	TDT-016
Width	mm	2,664	2,990	3,340
High	mm	5,460	5,770	5,950
Diameter	mm	2,500	2,800	3,150
Power	kW	45	75	110
Speed	r.p.m.	120	115	110
D. Turbine	mm	1,080	1,160	1,300
Capacity	m^3	7	10	16
Weigh	kg	6,950	9,050	13,200
Weigh full load	kg	17,450	24,050	37,200

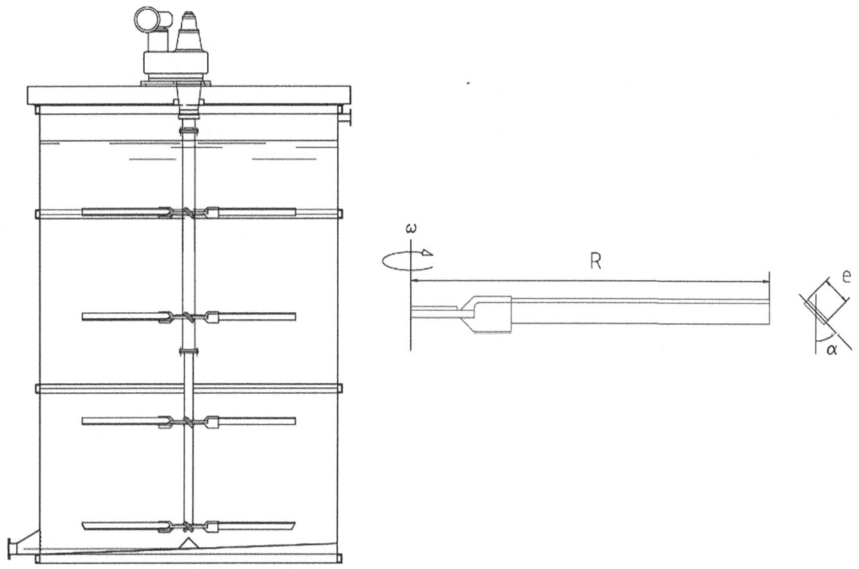

Fig. 4.7 Tank for stock of slip

The mechanical torque to move the blades can be estimated from the fluid dynamics of the relative flow:

$$T = C_D \cdot \rho \cdot \omega^2 \cdot e \cdot \sin\alpha \cdot \frac{R^4}{4} \tag{4.14}$$

where:

- T: Torque of rotation (N•m)
- C_D: Drag coefficient of the rectangular blade = 2
- ρ: Density of the fluid = 1,800 (kg/m^3)
- ω: Rotational speed = 40 (r.p.m.)
- e: Width of the blade = 0.1 (m)
- α: Torsional angle of the blade = 25 (deg)
- R: Total length of the blade = 1 (m).

Thus, rotational power yields:

$$N = T \cdot \omega(\text{kW}). \tag{4.15}$$

This is the power to move just one blade. It has to be multiplied by the total number of blades

Other electrical equipment within this section is run by smaller motors in the range of some kW's for belts, hoppers, etc. They may be simulated as generic electrical model in Sect. 3.2. Figure 4.8 shows the general scheme for this section.

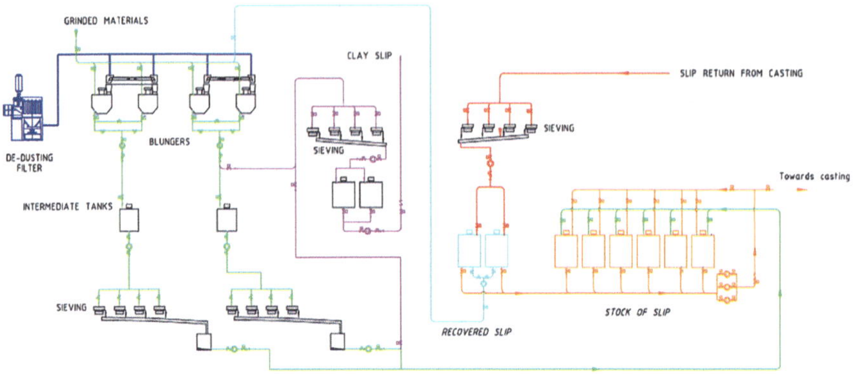

Fig. 4.8 Slip preparation general scheme

4.3 Glaze Preparation

The mathematical model of the glaze preparation section can be scaled from the slip preparation. The nature of the raw materials and the process are essentially the same. Table 4.10 shows a formula of raw material for glaze as an example.

For this formula and a final glaze density of 1,800 kg/m^3, Table 4.11 shows ratio of solids/water for the glaze liquid.

Known the particle size before and after milling, the energy to carry it out can be calculated from (4.3). Thus, the energy to mill the hard solids can be calculated as:

Then, the required electrical power to mill is:

Table 4.10 Formula and density of glaze raw materials [1]

Glaze formula			rho kg/m^3
Plastic materials	Kaolin	10.0%	2,948
Hard materials	Feldspar	29.0%	2,848
	Nepheline	5.0%	2,600
	Quartz	20.0%	2,650
	Calcium carbonate	9.0%	2,710
	Wollastonite	7.0%	2,950
	Dolomite	5.0%	2,840
	Magnesium carbonate	2.0%	2,960
	Barium carbonate	1.0%	4,290
	Talcum	2.0%	2,750
	Zinc oxide	2.0%	5,610
	Zirconium silicate	8.0%	4,560
	Total=	100%	3,007

Table 4.11 Water/solids ratio for prepared glaze

Glaze	Density	Mass	Volume
Water/solids ratio	kg/m^3	kg	m^3
Glaze	1,850	8,828	4.77
Water	1,000	2,842	2.84
Solids	3,007	5,987	1.99

$$N = \frac{E_{\text{hard solids}}}{t_{\text{cycle}}}(\text{kW}). \tag{4.16}$$

And the total milling volume is calculated dividing the slip volume by 0.45 as per Fig. 4.4:

$$m^3_{\text{workingvolume}} = \frac{m^3_{\text{glaze}}}{0.45}. \tag{4.17}$$

From Table 4.7, it can be calculated the proper size of the mill and from Table 4.9 the size of the blunger.

There is as well a need for storage with great spare of prepared glaze. Glaze tanks use to be smaller than slip.

4.4 Modeling and Molds

Yet the equipment is different for both types of factory, the electrical consumption at this section is not very relevant, either in size or number of motors. Thus, they may be considered as generic electrical model defined in Sect. 3.2.

4.4.1 Molds Dryer Model

The important consumption in the mold-making section of a plaster mold factory is the thermal requirement to evaporate water from molds after cast them. This point represents the first difference about consumptions between both types of factories. For the resin factory, there is no water to be evaporated and consequently, this thermal consumption does not account.

A mold dryer is basically a chamber where the water of the cast mold is slowly evaporated in a controlled environment where temperature and humidity have to be kept within a defined range of values. The operation parameters are shown in Table 4.12.

X represents the water contained in the air (Fig. 4.9).

Table 4.12 Operation parameters of molds dryer

Air condition	T^a	HR	e	$x_{\text{saturation}}$	x
	°C	%	mbar	g/kg	g/kg
Ambient	25	55%	31.69	20.08	11.05
Dryer chamber	60	75%	199.37	152.42	114.31

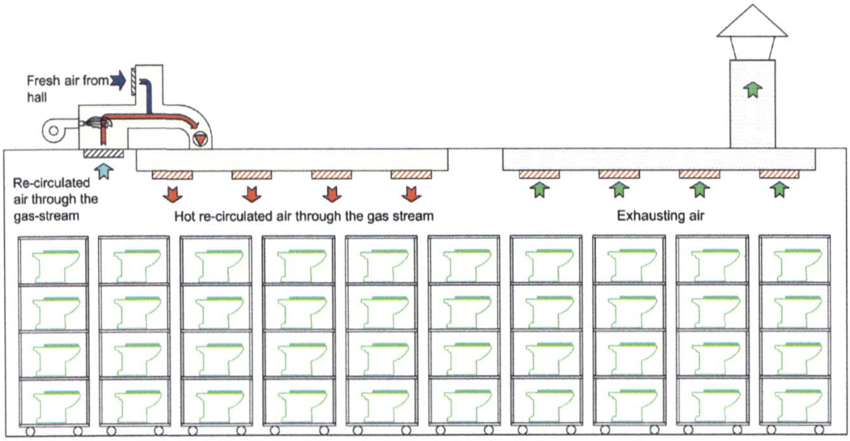

Fig. 4.9 Descriptive scheme of molds dryer (*Source* Data compiled by the authors)

The process inside the dryer consists of maintaining the psychrometric conditions inside by recirculating air with eventual thermal input. A small flow of fresh, relatively dry air, is renovated through the ventilation system in order to release the moisture.

The procedure to define the model is to find the following values:

- daily production of molds
- water contained in one mold to be evaporated
- fresh air and psychrometric conditions to absorb evaporated water
- heat losses through the dryer enclosure
- required energy to heat up all masses inside the dryer

4.4.1.1 Daily Production of Molds

The average life of a plaster mold is about 80 casting cycles. Thus, the required amount of molds for the whole year should be:

$$n_{\text{molds}} = \frac{P_{\text{year}}}{P_{\text{mold}}} \tag{4.18}$$

where:

Table 4.13 Construction parameters of the dryer modeled

Dryer parameters		
Length	6	m
Width	6	m
High	4	m
Walls surface	168	m^2
Dryer volume	144	m^3
"U" Thermal transfer coefficient	2.83	W/m^2K
Cycle	7	days
Nº molds inside	80	molds
Supports mass inside	102	kg
Specific heat coefficient	0.522	kJ/kg/K

P_{year}: Production of one year (pieces)

P_{mold}: production of one mold (pieces/mold).

The desired quality of the dried mold to meet hygroscopic specifications requires a slow drying process that usually takes up to 7 days.

They must be arranged in batches to be dried. So, dryer volume has to be designed for:

$$n_{\text{moldsindryer}} = \frac{n_{\text{molds}}}{n_{\text{workingdays}}} \cdot t_{\text{cycle}} \tag{4.19}$$

where:

$n_{\text{molds in dryer}}$: number of molds to be dried per cycle

$n_{\text{working days}}$: number of working days through the whole year

t_{cycle}: cycle time (days).

Construction parameters for the dryer model selected for this example are shown in Table 4.13:

Three units are required to operate this process with some spare.

4.4.1.2 Water Contained in One Mold to Be Evaporated

Table 4.14 shows the parameters that define the mold.

From these values, it can be calculated the water to be evaporated in one cycle:

$$m_{\text{evaporate}} = V_{\text{mold}} \cdot r_{w/p} \cdot \rho_{\text{mold}} \cdot n_{\text{molds in dryer}} \tag{4.20}$$

where:

$m_{\text{evaporate}}$: mass of water contained in the mold to be dried (kg)

Table 4.14 Mold parameters

Mold parameters		
Mold volume	0.231	m^3
Specific heat of the mold	0.8	kJ/kg/K
Ratio water/plaster	1.3256	kg_w/kg_p
Water/plaster mixture density	2,600	kg/m^3
Dried mold density	991	kg/m^3
Mold mass	229	kg
Porosity	61.88%	Capillarity inside the mold

V_{mold}: volume of the mold (m^3)

$r_{w/p}$: ratio of masses water plaster

ρ_{mold}: dried mold density (kg/m^3)

$n_{molds\ in\ dryer}$: number of molds to be dried per cycle.

Hence, the quantity of water to be evaporated in every cycle would be:

4.4.1.3 Fresh Air and Psychrometric Conditions to Absorb Evaporated Water

From Table 4.12 and the amount of water to be evaporated, it can be obtained the fresh air to absorb the evaporated water by:

$$m_{air} = m_{evaporate} \cdot \frac{1}{(x_{chamber} - x_{ambient})} \tag{4.21}$$

where:

m_{air}: mass of psychrometric air (kg)

$m_{evaporate}$: mass of water to evaporate (kg)

$x_{chamber}$: water content in the chamber environment (kg_{water}/kg_{air})

$x_{ambient}$: water content in the ambient air (kg_{water}/kg_{air}).

4.4.1.4 Heat Losses Through the Dryer Enclosure

A simple approach of heat transfers through the walls and roof of the drier provides the estimations of thermal losses during the process as per (4.22)

$$Q_{\text{losses}} = U \cdot S_{\text{enclosure}} \cdot \Delta T \tag{4.22}$$

4.4.1.5 Required Energy to Heat up All Masses Inside the Dryer

By means of (4.23), they can be calculated the required thermal energy to heat up supports and molds as solids. On the other hand, (4.24) provides the inlet and outlet enthalpies for:

- The vaporized water from molds
- Fresh air to absorb the vapor
- Natural gas (assumed as CH_4), and combustion air inlets
- CO_2, H_2O from combustion, N_2 and excess air outlets as mixed exhausting flow

and from these values can be derived the required heat to be supplied by the amount of natural gas.

Mass and energy balances yield the amount of fuel required (Fig. 4.10).

$$\sum m_{\text{IN}} = \sum m_{\text{OUT}} \tag{4.23}$$

$$\sum H_{\text{IN}} = \sum H_{\text{OUT}}. \tag{4.24}$$

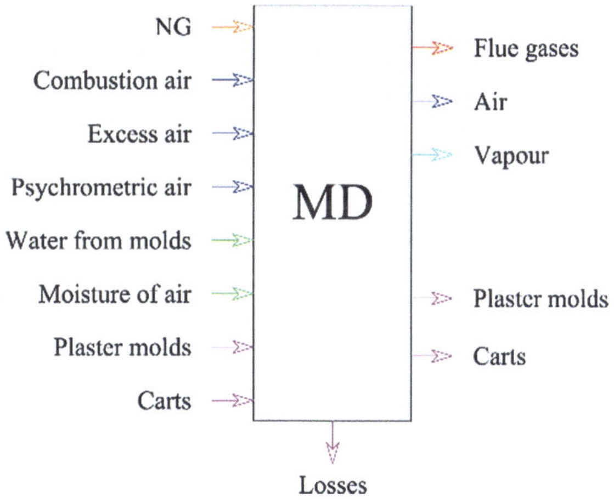

Fig. 4.10 *Model scheme of a mold's dryer* (*Source* Data compiled by the authors)

4.5 Casting

The casting section is characterized essentially by the molds and system to fill them with the barbotine, keep a pressure during the cycle and finally empty them of barbotine excess. There are involved some electrical equipment as; pumps, motors. Compressed air is required to act pneumatic devices for movement and automation as well as blowing molds when releasing the cast pieces. On the other hand, there are some elements that are required for both types as; slip tanks, ponds for return slip, pumps.

In essence, the difference lies in the machinery directly linked to molds.

4.5.1 Plaster Mold Factory

This section defines the main difference between the two factory models. On one side, for plaster molds, long benches are required to dispose multiple molds with two casts per day capacity. Thus, the surface for this section is quite larger in this type of factory. Figure 4.11, shows an example of this kind of benches.

Fig. 4.11 *Casting bench for toilet seats*

4.5.2 Resin Mold Factory

On the other hand, a resin mold factory requires pressure machines where the molds are able to cast continuously in cycles varying from some minutes to an hour depending on the size and geometry of the piece. Figure 4.12 shows the scheme of a pressure machine for toilet seats. It can be seen how the molds are pressed each other by means of a cylinder pushed by an oil pressure system. About 35 Nm^3 per cycle are required to act automation devices and blown off the cast piece.

4.5.2.1 Boilers

The water to flush molds after every cast must be warmed previously up to 45 °C from the standard state at 25 °C. The mathematical model starts from the water amount which is based on the manufacturer's specifications to properly clean the molds. Every mold requires 0.18 m^3 after every cycle. Thus, the water flow to be warmed up is:

$$m_w = \frac{P_{\text{year}}}{n_{\text{workingdays}}} \cdot \frac{1\text{ year}}{232\text{days}} \cdot \frac{1\text{day}}{24\text{hours}} \cdot \frac{1\text{hour}}{3,600\text{s}} \cdot V_{\text{mold}} \cdot \rho_w \tag{4.25}$$

where:

m_w: mass flow of water to be warmed (kg/s)

P_{year}: production of one year (pieces)

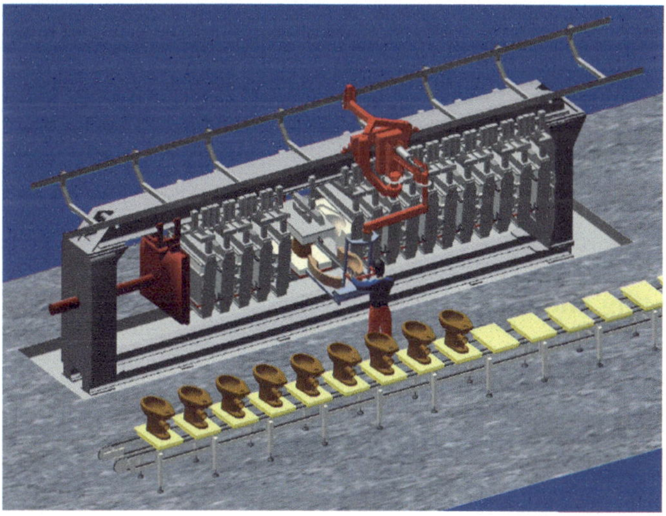

Fig. 4.12 *Pressure machine for WC casting [2]*

Fig. 4.13 *Thermal model of a conventional boiler* (*Source* Data compiled by the authors)

$n_{\text{working days}}$: number of working days in one year (days)

V_{mold}: volume of water to flush one mold (m^3)

ρ_w: density of the water (kg/m^3).

And the thermal power required:

$$Q = m_w \cdot (h_{\text{out}} - h_{\text{in}}) \tag{4.26}$$

where:

Q: thermal power to warm water (kW)

m_w: mass flow to be warmed up (kg/s)

h_{out}: enthalpy of warmed water (kJ/kg)

h_{in}: enthalpy of water inlet (kJ/kg).

xxx shows the scheme of the thermal model for a conventional boiler (Fig. 4.13). From (3.1) to (3.2), the system of equations to be solved is:

$$m_{NG} + m_{\text{air}} + m_{\text{water}} = m_{\text{fluegases}} + m_{\text{water}} \tag{4.27}$$

$$\begin{aligned} m_{NG} \cdot h_{NG} + m_{\text{air}\cdot h_{\text{air}}} + m_{\text{water}} \cdot h_{\text{water}} = m_{\text{fluegases}} \cdot h_{\text{fluegases}} \\ + m_{\text{water}} \cdot h_{\text{water}} + Q_{\text{losses}}. \end{aligned} \tag{4.28}$$

4.6 Heating and Ventilation

Electrical requirement is not very representative within this section and it has to be considered as a generic electrical model. Deriving from the differences in the cast technology, the thermal requirements for heating and ventilation are quite higher in a plaster mold factory due to two main factors:

- Greater surface to be dealt with
- The amount of water to be evaporated from plaster molds

4.6.1 Heating and Ventilation Unit Model

In order to evaluate the evaporation thermal requirement, it has to be calculated the amount of water contained in the mold after casting.

Table 4.15 shows the amount of water contained in a slip or ware from slip preparation till drying for every piece of ware as the percentage of water relative to final dry weigh.

The procedure to remove water from a slip is sequentially described below:

- Water from slip to mold that has to be removed from the mold.
- Water evaporated from ware during casting environmental conditions.
- Water evaporated from both the ware and mold during mold drying environmental conditions.
- Water evaporated from the ware within the ware dryer (WD).

The amount of water to be removed during each process is shown in Table 4.16

Evaporation of water is carried out by means of controlled psychrometry within the environment relative to every process in either the casting hall or dryer chamber [3].

After the casting procedure, the ware is demolded and stood with no manipulation until up 20% of its water content has it is dried up. Once this point has been reached, the temperature and humidity setup is changed into drying conditions with higher temperature and lower humidity in order to evaporate the water in the molds. Meanwhile, water is reduced from 20% to 17% in the standing ware. After this period (next duty cycle), the mold is ready to work again with fresh slip and the ware to be taken into the dryer where the remaining 17% of moisture has to be removed.

Table 4.15 Water content for different processes [1]

Water content/Dry weigh	%	kg
Slip	48%	9.62
Demolding	21%	4.20
After casting period	20%	4.00
Entering the ware dryer	17%	3.40

Table 4.16 Water to be eliminated from ware during different processes per day

Water to be eliminated	kg/piece	kg/day
Water in slip	9.62	20,742
Water retained in mold	5.42	11,690
Water evaporated in HVU in casting period	0.20	431
Water evaporated in HVU in drying period	0.60	1,293
Water evaporated in Ware-Dryer	3.40	7,328
TOTAL=	9.62	20,742

At this point, known water to be evaporated, as well as drying ability of the environment, it can be computed the thermal requirement to carry out the evaporation besides thermal requirements for transmission losses and solar radiation.

In a casting hall, humidity and temperature must be monitored in order to control the environment during the casting and the mold drying. For this reason, this process may be simulated as if the whole hall is a dryer where both, pieces and molds are to be dried and the humid air has to be removed outdoors. The scheme of the unit is shown in Fig. 4.14.

For the purpose of water evaporation, the key variable directly linked to temperature and relative humidity is the amount of air needed to saturate the atmosphere.

Depending on the heating media and the process, inlets are as follows:

- NG (mainly CH_4)
- Air for combustion and excess air
- Air to dry the humidity by psychrometry
- Moisture in the air
- Water from ware and molds to be evaporated
- Ware and other solids (Carts)

The outlets are:

- Combustion products (mainly CO_2, H_2O, N_2)
- Air heated and saturated
- Vapor within the saturated air
- Ware and other solids (Carts)

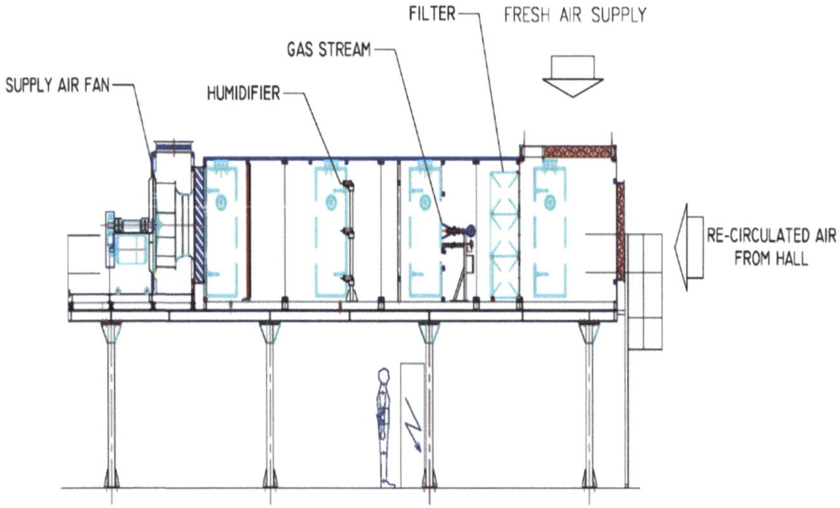

Fig. 4.14 HVU descriptive scheme

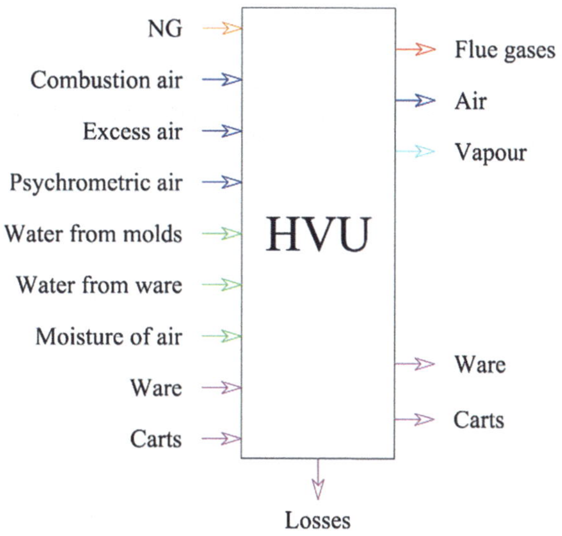

Fig. 4.15 Model scheme of a HVU

HVU process is regulated within specific ranges of temperature and humidity. When the process is led by NG combustion, the regulation of temperature is carried out by adjusting the NG supply and the humidity by fresh dry air. The system of (3.2)–(3.18) is applied to the model for every single period along the whole year in order to establish the values of the different variables on each stage of operation or climate season. The thermodynamic model is shown in Fig. 4.15.

4.6.2 Thermal Demand Calculation of a HVU

The aim of the process is to evaporate the water contained in the ware and mold by blowing fresh air relatively dry and warm into the casting hall, which is maintained in the specified hygrothermal conditions. Fresh air is taken from outside the factory and mixed with dry air from inside depending on both hygrothermal states.

The thermal requirement comes from:

- heat carts and ware
- heat the air to carry the evaporated water
- thermal losses throughout the enclosure
- solar radiation that helps heating reduction
- evaporation of the water to be extracted

The total thermal requirement for the system is:

$$Q_{\text{thermal}} = Q_{\text{carts\&ware}} + Q_{\text{ventilation}} + Q_{\text{losses}} - Q_{\text{solar}} + Q_{\text{evaporation}}. \tag{4.29}$$

Calculation of each term is detailed as follows.

4.6.2.1 Heat up the Ware and the Supporting Carts

$$Q_{\text{carts\&ware}} = (m_{\text{ware}} \cdot Ce_{\text{ware}} + m_{\text{carts}} \cdot Ce_{\text{carts}}) \cdot \Delta t. \tag{4.30}$$

where:

m_{ware} = mass of ware in the chamber

Ce_{ware} = specific heat for the ware

m_{ware} = mass of carts in the chamber

Ce_{ware} = specific heat for the carts

Δt = increase of temperature as per each stage of drying process.

4.6.2.2 Heat the Fresh Air to Absorb Evaporated Water

$$Q_{\text{ventilation}} = m_{\text{freshair}} \cdot Ce_{\text{air}} \cdot \Delta t \tag{4.31}$$

where:

m_{freshair} = mass of fresh air needed to evaporate

Ce_{air} = specific heat for the air at the required temperature

Δt = temperature difference between inside and outside of the hall.

The needed amount of fresh air taken from either the hall or external is calculated by:

$$m_{\text{freshair}} = \frac{m_{\text{evaporation}}}{x_{\text{in}} - x_{\text{out}}} \tag{4.32}$$

where:

$m_{\text{evaporation}}$ = mass of water to be evaporated in each stage of the process. It can be taken from Table 4.16.

x_{in} = moisture saturation of air in the hall

x_{out} = moisture of air at external condition.

4.6.2.3 Keep the Temperature into the Chamber Because of the Thermal Losses

$$Q_{\text{losses}} = S \cdot U \cdot \Delta t \tag{4.33}$$

where:

S = surface of chamber enclosure

U = global coefficient of heat transmission

Δt = temperature difference between inside and outside.

4.6.2.4 Keep the Temperature into the Chamber Because of the Solar Radiation

$$Q_{\text{radiation}} = \varepsilon \cdot \sigma \cdot T_{\text{roof}}^4 \tag{4.34}$$

where:

ε = emissivity of the tin on the roof

σ = Stephan-Botzman's constant ($5.67 \bullet 10^{-8}$ ($W/m^2/K^4$))

T_{roof} = temperature of the roof due to radiation.

When irradiated, the temperature of the roof increases up to the point to equal the temperature inside. At this moment, the conduction is stopped as well as thermal losses and the balance of them becomes negative.

4.6.2.5 Evaporate the Water to Be Extracted

$$Q_{\text{evaporation}} = m_{\text{evaporation}} \cdot \left(h_{\text{vapor}} - h_{\text{water}}\right) \tag{4.35}$$

where:

h_{vapor} = specific enthalpy of vapor at hall condition

h_{water} = specific enthalpy of water at hall condition.

4.6.3 Climate Conditions

Climate variability affects HVU performance all year long because of the temperature, humidity and solar radiation. Available data allow to take this effect into account by calculating thermal performance every half hour.

In order to absorb water from the ware and mold, the air has to be unsaturated but within the ranged temperature and humidity that are shown in Table 4.17 for both types of factory depending on the casting technology.

Environmental drying performance of the air depends on the psychrometric characteristics of the ambient air. For the case study of this book, the temperature, humidity and solar radiation provided by National Weather Service for 1 year in periods of 30 min are shown in Fig. 4.16.

Table 4.17 Parameters of HVU cycle

States		PMF		RMF
		Casting	Drying	Casting
Time	h	10	14	24
T_{hall}	°C	25	40	25
T_{hall}	K	298	313	298
x_{hall}	%	75%	50%	75%
$e_{w\ \text{hall}}$	Pa	31.69	73.77	31.69
$x_{s\ \text{hall}}$	g/kg	20.08	48.86	20.08
x_{hall}	g/kg	15.06	24.43	15.06
T_{external}	°C	16.86	13.96	15.17
T_{external}	K	290	287	288
x_{external}	%	61.29%	72.07%	67.58%
$e_{w\ \text{external}}$	Pa	19.21	15.95	17.24
$x_{s\ \text{external}}$	g/kg	12.02	9.95	10.77
x_{external}	g/kg	7.37	7.17	7.28
Initial ware water content	kg	1,724	1,006	1,724
Water to evaporate from ware	kg	718	1,006	1,724
Final ware water content	kg	1,006		
Initial mold water content	kg		11,690	
Water to evaporate from mold	kg		11,690	

It can be seen as general trend that:

- External temperature grows by the middle of the day and, obviously, it is greater in summer
- Regarding humidity, it is lower at noon and grows at night
- Radiation, obviously grows at noon. This fact helps provide a decrease of thermal consumption.

4.7 Drying

The focal point of this section from the perspective of consumptions is the dryer. Some other equipment may be involved, especially for conveying ware in and out. Nevertheless, as electrical equipment it can be considered within the electrical generic model.

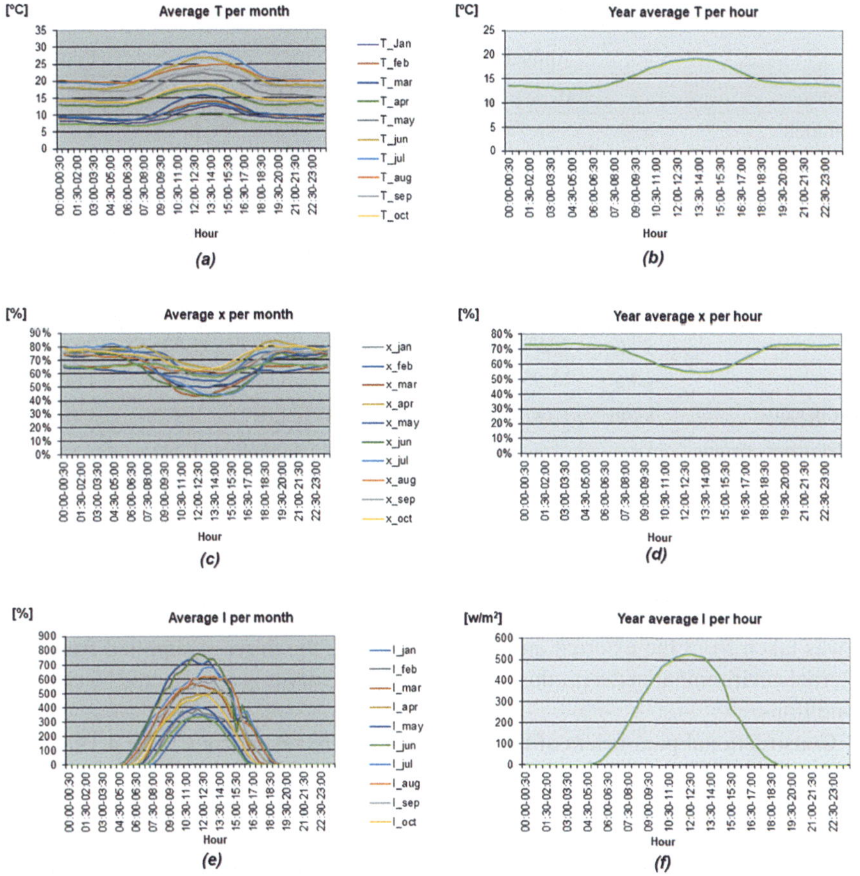

Fig. 4.16 Average climate conditions per hour of the day (T for Temperature, x for humidity and I for Solar radiation). **a**, **c** and **e** represent the averaged values of every month. **b**, **d** and **f** the annual averages

4.7.1 Ware Dryer

From the electrical point of view, a ware dryer can be considered as a generic electrical model. Regarding the drying process, climate changes do not affect performance since the fresh air required is taken from the factory hall, which has previously treated by HVU.

According to Table 4.15, the ware is taken with 17% of water content into the dryer whose environmental conditions are shown in Table 4.18 for the five drying stages.

The process in the dryer is, from a thermodynamic point of view, the same as in a HVU. Fresh air and warm air are blown through the ware in order to evaporate the water inside. The main difference with HVU is the source of fresh air. While

Table 4.18 Operation parameters of ware-dryer cycle

		Ambient	Heating				Cooling
Stages		0	1	2	3	4	5
Duration	h		2	3	2	3	1
Time	h	0	2	5	7	10	11
T	°C	25	38	50	75	90	25
T	K	298	311	323	348	363	298
RH	%	55%	75%	50%	30%	20%	55%
e	mbar	31.69	64.49	123.39	386.54	705.29	31.69
x_s	g/kg	20.08	42.29	86.28	383.79	1,425.68	20.08
x	g/kg	11.05	31.72	43.14	115.14	285.14	11.05
Air density	kg/m^3	1.1818	1.1343	1.0904	1.0121	0.9703	1.1818
Inlet water content	kg		3,664	3,125	2,371	1,293	
Water to evaporate	%		2.50%	3.5%	5.00%	6.00%	
Outlet water content	kg		3,125	2,371	1,293	0	
Water to evaporate	kg		539	754	1,078	1,293	

it was taken from the external atmosphere to HVU, the dryer is supplied from the internal environment in the production hall. Thus, the dryer is not affected by climate conditions.

Constructional parameters of the ware dryer can be taken from Table 4.19.

The thermal power of the dryers is basically dedicated to evaporate water from the pieces and to heat them up to the designed temperature in every stage of the drying process compensating the heat losses throughout the insulated enclosure. Besides, the constant flow of air needed to evaporate the water requires additional heat to reach the designed temperature inside the dryer. The dryer operation scheme is shown in Fig. 4.17.

The process is carried out in different stages in order to keep the flow of water out from the core of the pieces. When the cycle is starting, humidity in the chamber is critical to avoid a draught onto the surface of the ware that would cut the water way out. Later on, the temperature is increased and humidity is reduced, when most of the internal water has already reach the surface. The process is very similar to the heating and ventilation one. The only difference lies in the mold water, which in this case does not make sense, since there are no molds into the ware dryer. On the other hand, as said before, climate conditions do not affect it since the surrounding environment is the hall of the factory.

The thermodynamic model is shown in Fig. 4.18.

Table 4.19 Construction parameters of the ware dryer

Production parameters of the ware dryer		
Length	21,520	mm
Wide	11,800	mm
High	2,900	mm
"U" enclosure	2.83	W/m^2/°C
Heating transfer surface	332	m2
Carts number	64	
Pieces per cart	16	
Cart weigh	200	kg
Production	5.00E + 05	Pieces/year
Labor days	232	days/year
Nª cycles	2	cycles/day
Pieces in dryer	1,078	pieces in dryer
Wet weigh of piece	23	kg
Dry weigh of piece	20	kg
Initial water content	17.00%	Wet base
Water content	3,665	kg
Specific heat of cart	0.522	kJ/kg/°C
Specific heat of ware	1.13	kJ/kg/°C
LHV NG	48,000	kJ/kg
Temperature of the hall	25	°C
Humidity in the hall	50.00%	

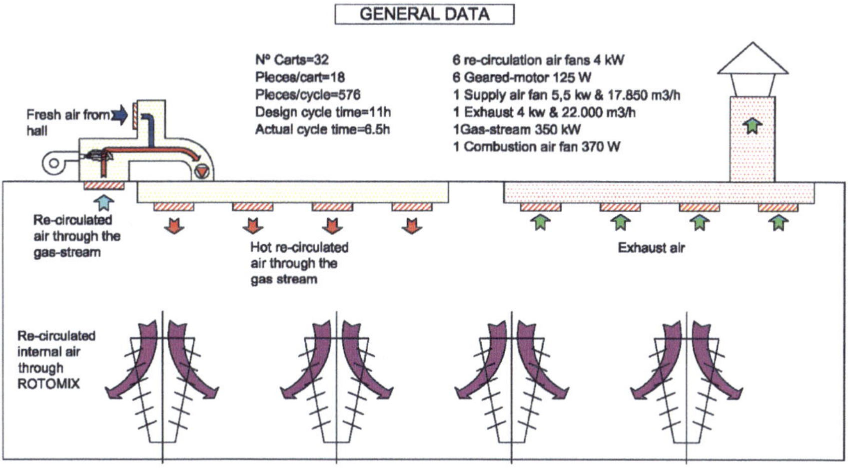

Fig. 4.17 Operation scheme of ware dryer

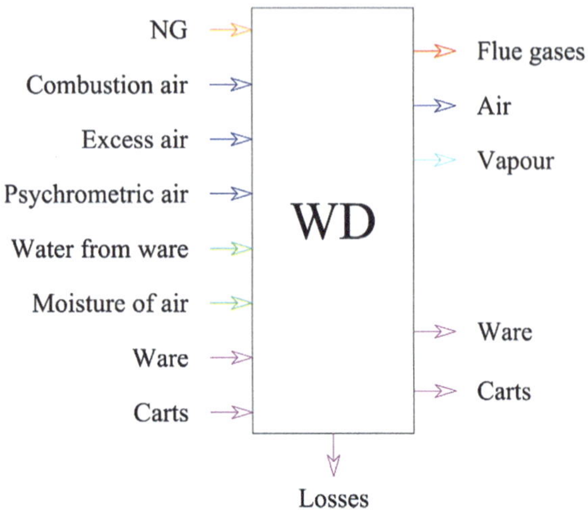

Fig. 4.18 Model scheme of a ware dryer

4.7.2 Thermal Demand Calculation of a Ware Dryer

The aim of the process is to evaporate the water contained in the ware by blowing fresh air relatively dry and warm into the casting hall, which is maintained in the specified hygrothermal conditions. Fresh air is taken from the hall of the factory.

The thermal requirement comes from the need to:

4.7.2.1 Heat up the Ware and the Supporting Carts

$$Q_{\text{carts\&ware}} = (m_{\text{ware}} \cdot Ce_{\text{ware}} + m_{\text{carts}} \cdot Ce_{\text{carts}}) \cdot \Delta t \tag{4.36}$$

where:

m_{ware} = mass of ware in the chamber

Ce_{ware} = specific heat for the ware

m_{ware} = mass of carts in the chamber

Ce_{ware} = specific heat for the carts

Δt = increase of temperature as per each stage of drying process.

4.7.2.2 Heat the Fresh Air to Absorb Evaporated Water

$$Q_{\text{ventilation}} = m_{\text{fresh air}} \cdot Ce_{\text{air}} \cdot \Delta t \tag{4.37}$$

where:

$m_{\text{fresh air}}$ = mass of fresh air needed to evaporate

Ce_{air} = specific heat for the air at required temperature

Δt = temperature difference between hall and exterior.

The needed amount of fresh air taken from either the hall or external is calculated by:

$$m_{\text{fresh air}} = \frac{m_{\text{evaporation}}}{x_{\text{IN}} - x_{\text{OUT}}} \tag{4.38}$$

where:

$m_{\text{evaporation}}$ = mass of water to be evaporated in each stage of the process. It can be taken from Table 4.16.

x_{IN} = moisture saturation of air inside the chamber

x_{OUT} = moisture of air in the hall.

4.7.2.3 Keep the Temperature into the Chamber Because of the Thermal Losses

$$Q_{\text{losses}} = S \cdot U \cdot \Delta t \tag{4.39}$$

where:

S = surface of chamber enclosure

U = global coefficient of heat transmission

Δt = temperature difference between inside and outside.

4.7.2.4 Evaporate the Water to Be Extracted

$$Q_{\text{evaporation}} = m_{\text{evaporation}} \cdot \left(h_{\text{vapor}} - h_{\text{water}}\right) \tag{4.40}$$

where:

h_{vapor} = specific enthalpy of vapor at dryer condition

h_{water} = specific enthalpy of water at dryer condition.

The total thermal requirement for the system is:

$$Q_{\text{thermal}} = Q_{\text{carts\&ware}} + Q_{\text{ventilation}} + Q_{\text{losses}} + Q_{\text{evaporation}}. \tag{4.41}$$

4.8 Glazing

Glazing section does not require thermal equipment. Electrical and pneumatic devices provide the power to move, automate, illuminate the involved machinery. Thus, electrical generic model is applied to any of them as stated in Sect. 3.2.1.

4.9 Kilns

Electrical requirement for this section is modeled as a generic electrical model.

Thermal consumption account for most of the energy consumed throughout the whole global process. It may reach over 50% of the total energy consumption. Therefore, they must be analyzed in detail as the main source of potential savings along the factory.

4.9.1 Tunnel Kiln Model

Tunnel kiln configuration, model and calculation are described in [4]. For the purpose of the case study in this book, the model scheme of the kiln has been considered as follows in Fig. 4.19.

As the tunnel kiln is running on a steady basis [5], it is convenient to consider taking advantage of its residual heat. The great amount of thermal power that flows through it is another reason to assess the energy flow. In brief, the kiln is run as a burner of natural gas along an insulated tunnel where air flow is run upward in order

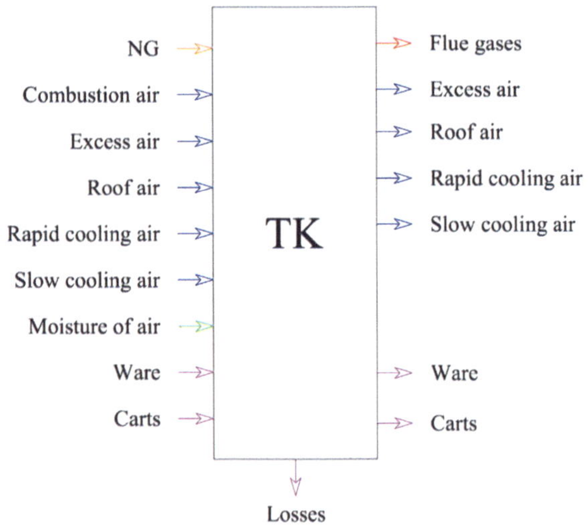

Fig. 4.19 Thermal model scheme of tunnel kiln

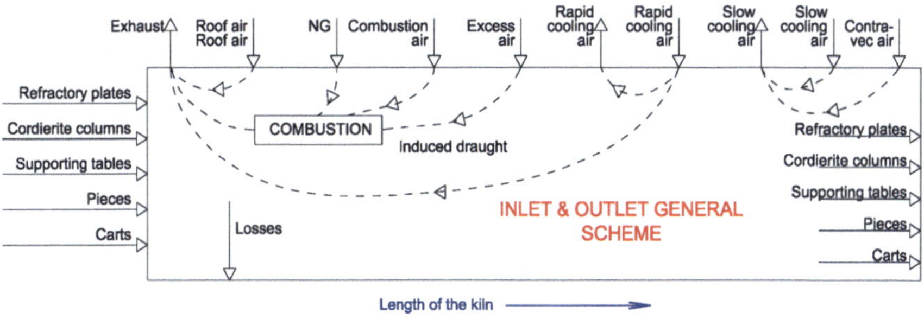

Fig. 4.20 Tunnel kiln scheme of flows

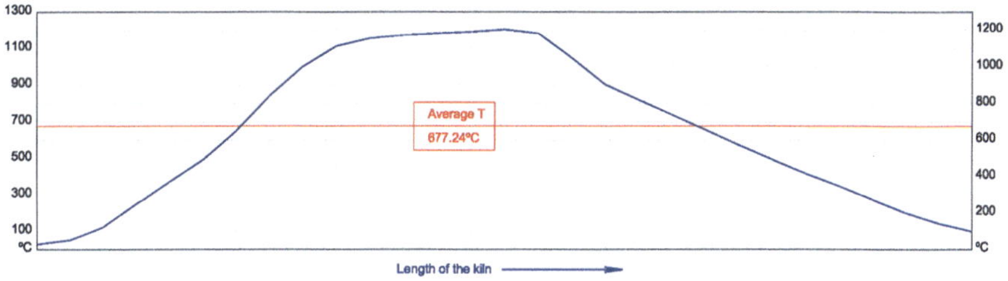

Fig. 4.21 Temperature distribution along the kiln

to bring the ceramic pieces up to the point of vitrifying the glaze which is around 1250 °C. Three different air streams are led through it for cooling and firing the pieces as shown in Fig. 4.20

The thermal power is basically devoted to heat ware, carts, supports, superstructure and auxiliary air flow up to designed temperature along the tunnel. All the heat capacity after the top of the curve (at about 1250 °C) (see Fig. 4.21) is delivered downstream to cooling airs. On the other hand, a significant quantity of heat is lost throughout the insulation enclosures, which mainly consist of refractory bricks and fiber liners.

As any other thermal consumer, the kiln is resolved through mass and energy balances. Nevertheless, the configuration is a bit more complex and so does the system of equations.

The mass balance equals the air inlets (roof air, combustion and excess airs, rapid and slow cooling airs), the natural gas mass inlet, air outlets (exhausted gases and exhausted cooling airs) and load.

$$\sum \dot{m}_{\text{in}} = \sum \dot{m}_{\text{out}}. \tag{4.42}$$

On the other hand, the energy balance equals the thermal inlet energies and the sum of outlet energies and losses. Every term of the equation is calculated multiplying the mass flow by the enthalpy of it at the corresponding temperature:

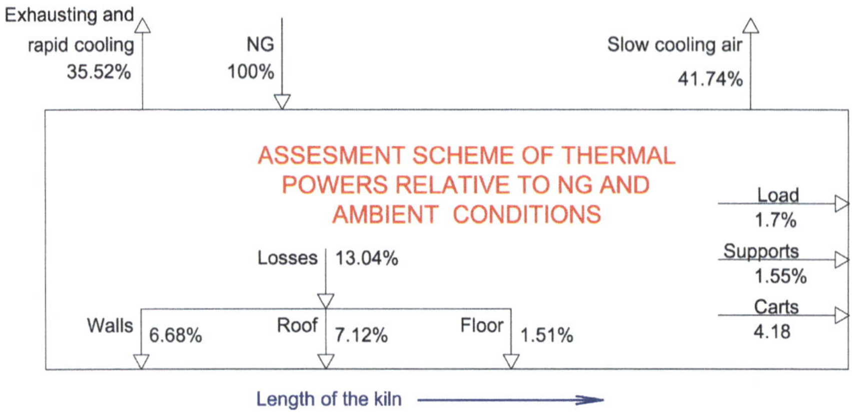

Fig. 4.22 Energy balance in a tunnel kiln

$$\sum \dot{m}_{\text{in}} h_{\text{in}} = \sum \dot{m}_{\text{out}} h_{\text{out}} = +W_{\text{losses}}. \tag{4.43}$$

Considered energy inlets are:

- Roof, combustion, excess, rapid cooling, slow cooling airs.
- Natural gas flow rated at 48,000 kJ/kg.
- Cars and superstructure, consisting of: ceramic pieces, supports, tables and plates.

Losses are calculated considering thermal transference from inside the kiln to outside including the three components of the heat transference; conduction, convection and radiation.

Yet the model is based on a simple set of concepts, the solution is complex due to the fact that temperature varies along the tunnel so as enclosure configuration and blown air flows.

Figure 4.22 shows the net energy balance through the different sections of the tunnel.

4.9.1.1 Kiln Definition

Accordingly with the best available technologies for sanitary-ware manufacturing [6] at 2.3.7.5 regarding the firing of the ware, the most common way to fire sanitary ware is through tunnel kilns which essentially, are refractory tunnels served by rail tracks carrying kiln-cars. Figure 4.23 shows a standard tunnel for sanitary ware. The air and exhausting conducts are remarked. The kiln is constructed by adding 3 m long modules. The different modules for each section of the kiln can be seen.

Specifications for a standard tunnel kiln of the leader of sanitary ware machinery manufacturer are as follows in Fig. 4.24.

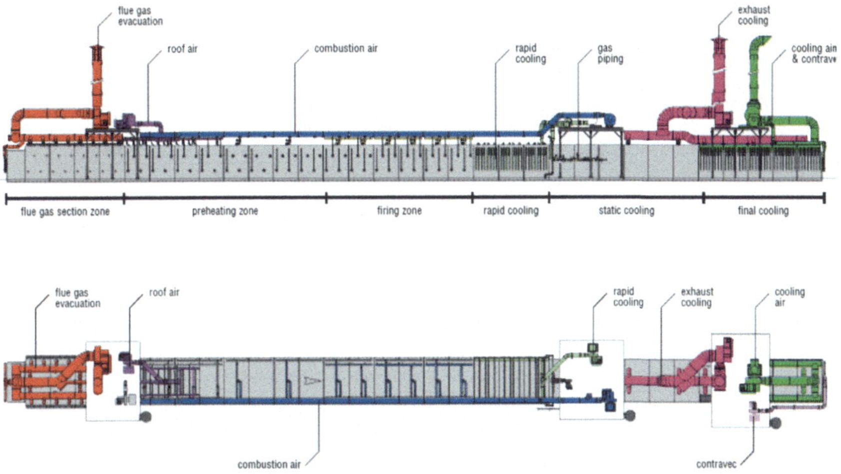

Fig. 4.23 Tunnel kiln configuration (*Source* SACMI Catalogue)

Fig. 4.24 Tunnel kiln technical specifications (*Source* SACMI Catalogue)

TECHNICAL DATA	
Kiln possible dimension	46 - 132 m
Useful width	2,400 - 3,800 mm
Possible car loading height	800 - 1,200 mm
Car useful volume	2.52 - 4.26 m^3
Car in the kiln	32 - 94
Average load on car (depending on products mix and loading conditions)	18 - 30 pieces
Maximun temperature	1,300 °C
Firing temperature	1,269 °C
Approximate firing cycle	10 - 20 hours
Approximate specific net consumption (depending on technological and firing cycle conditions)	4,180 - 5,852 kJ/kg

The kiln length chosen in this case study is 84 m, with 28 modules. The characteristics of such a kiln are listed in Table 4.20.

The carts have refractory decks on which, dried ware is set in defined stable patterns. Figure 4.25 shows a cross-section of this tunnel with a standard load configuration of sanitary-ware toilet seats. The carts are pushed through the kiln at set intervals. On the other hand, exhausting fan at the entrance of the tunnel produces a counter flow of induced draught of gases allowing a better heat exchange between gases and ware. The high temperatures inside the kiln are kept by high performance refractory materials with appropriate thickness.

Table 4.20 Main characteristics of a kiln 84 m. long (*Source* SACMI—Technical specifications)

Kiln specifications		
Kiln length	84	m
Kiln width	3	m
Firing temperature	1214	°C
Number of cart inside the kiln	62.2	
Duty cycle	16	h
Excess of air	10%	
Roof air	0.3120	kg/s
Rapid cooling air	0.7926	kg/s
Slow cooling air	1.6760	kg/s
Contravec air	1.4339	kg/s
Annual production	500,000	Pieces/year
Efficiency	86.54%	
Specific consumption	5,225	kJ/kg
Average mass of piece	20	kg/piece
Days/year kiln operation	325	days/year
Kiln thermal power	2,150	kW

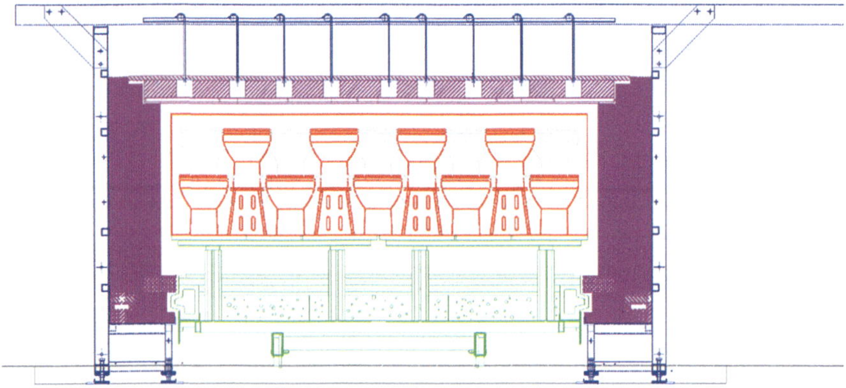

Fig. 4.25 Cross-section of the kiln car (Courtesy of SACMI GROUP)

The principle of operation is to heat up the load from inlet by hot gas aspirated from the firing zone up to module 15 in the middle of the kiln at about 1250 °C where the chemical transformation of the ceramics takes place. Then the load and cart start to be cooled by means of rapid cooling firstly and then slow cooling up to the kiln outlet. The thermal energy that load and cart carry when entering the cooling section is transferred to cooling air and sent out through stacks 2 (rapid cooling) and 3 (slow cooling) where they are collected to be recovered at the MED heating. Due to induced draught toward exhausting stack, it usually happens that rapid cooling air

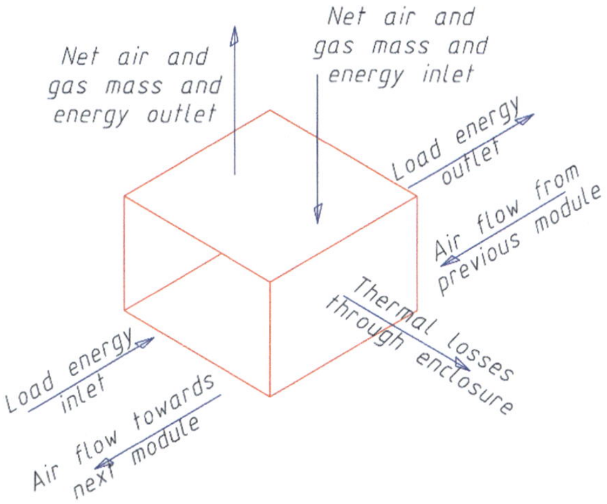

Fig. 4.26 Schematic mass and energy balances for each module

flow results inward rather than outward. Cooling air also improves the heat exchange along the kiln by means of a proper turbulence that may be optimized as stated in [7]. In [9], Mezquita studies how to reduce consumption of gas in the kiln by pre-heating the combustion air with recovered heat from cooling air.

4.9.1.2 Simulation Methodology

Simulation of tunnel kiln has been researched and some works have been carried out regarding the whole kiln [8], or temperature distribution [10], yet most of them are devoted to tile production rather than sanitary ware. A tunnel kiln is configured as an alignment of modules designed to operate under defined internal temperature. Hence, the composition of the wall and roof is variable with the longitudinal location.

In this case study, the model has been carried out by analyzing the mass and energy balances for each module independently and then mix all of them together in order to obtain the global balance for the whole kiln. For each module, it is calculated:

- the balance of air mass and thermal inlet and outlet air and gas flows.
- the balance of the energy of the load inlet and outlet, since the mass is not varying.
- the losses through enclosure walls, roof and floor by a combination of thermal conduction, convection and radiation.

Figure 4.26 shows the scheme for the balances.

4.9.1.3 Losses Calculation

The calculation of the losses is based on a conventional combination of radiation, convection and conduction heat transfer as Fig. 4.27 shows.

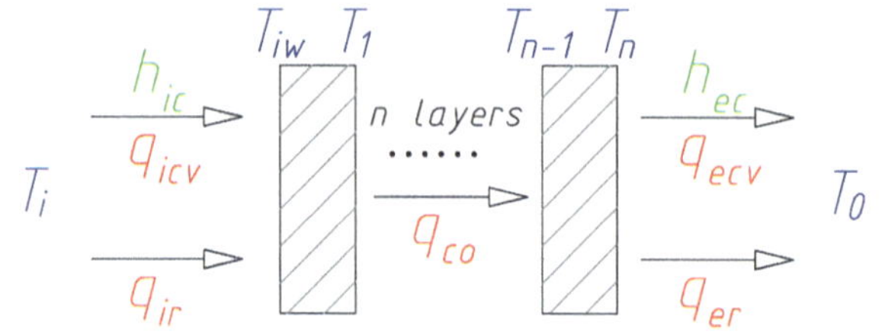

Fig. 4.27 Schematic path of thermal losses through the different insulation layers

The number and features of the layers depend on the section of the kiln. On the other hand, the performance of the insulation depends on the temperature. This fact makes the resolution more complicated, which has to be carried out by means of an iterative procedure to find out the temperature gradients and thermal flows.

As the temperature gradient varies along the kiln, so do the properties of the thermal insulation.

Figure 4.28 shows the configuration of masses that configure the cart.

Correspondent values to compute thermal transfer through them are shown in Table 4.21

The building configuration of the roof and side walls is carried out by different combinations of insulation layers depending on the kiln section. Heating and firing sections with higher temperatures must be strongly insulated with tough refractory materials. On the other hand, the cooling sections may be softened as the temperature gradient decreases as shown in Table 4.22.

Regarding the values of the thermal conductivity for each material, it has to be said that values are depending on the temperature as per manufacturer specifications. These values have been summarized in Table 4.23.

In order to determinate the actual conductivity at any temperature within the operation range, they have been adjusted by least square minimums to provide correspondent polynomial regressions as per Table 4.24.

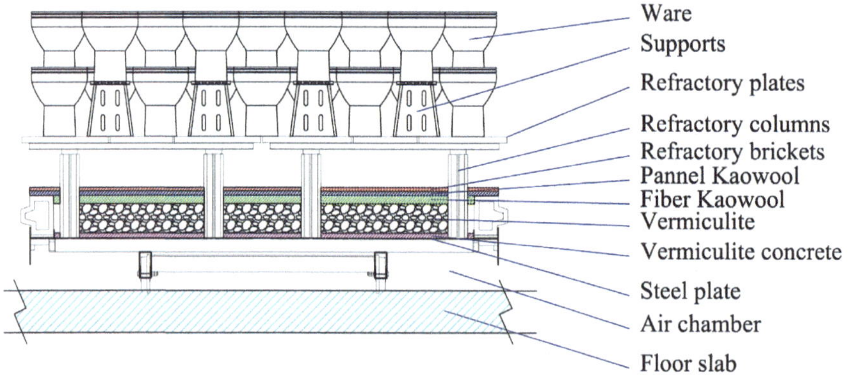

Fig. 4.28 Configuration of load and cart insulation

Table 4.21 Thickness and thermal conductivity for the insulation layers of the cart

Cart layers	Material	Thickness	l
		mm	W/K/m
Layer-1	Refractory concrete	20	0.60
Layer-2	Refractory panel	25	0.30
Layer-3	Refractory fiber	50	0.40
Layer-4	Vermiculite concrete 1/8	175	0.14
Layer-5	Vermiculite concrete 1/5	30	0.14
Layer-6	Steel plate	3	50.00
Air chamber	Air	315	0.026
Slab floor	Concrete	250	1.70

Polynomial is defined as:

$$\lambda\left[\frac{w}{mK}\right] = \sum C_i \cdot T^i. \tag{4.44}$$

Fiberglass conductivity is more complex to be determined because it depends simultaneously on the temperatures at both sides as shown in Table 4.25.

An exponential correlation has been adjusted by:

$$\lambda\left[\frac{w}{mK}\right] = C_1(T_{\text{cold}}) \cdot e^{C_2(T_{\text{cold}})\cdot T_{\text{hot}}}. \tag{4.45}$$

With C_1 and C_2 as polynomials functions of T_{cold} by:

$$C_1(T_{\text{cold}}) = \sum_0^7 a_i \cdot T_{\text{cold}}^i \tag{4.46}$$

$$C_2(T_{\text{cold}}) = \sum_0^7 a_i \cdot T_{cold}^i. \tag{4.47}$$

Polynomial coefficients result (Table 4.26).

Once defined the different configurations of insulation enclosures, the general set of equations to find out the losses so as the temperature gradient along the path is as follows, agreeing with Fig. 4.27.

$$q_{icv} = h_i \cdot S \cdot (T_i - T_1) \tag{4.48}$$

$$q_{ir} = \sigma \cdot \varepsilon \cdot S \cdot \left(T_i^4 - T_1^4\right) \tag{4.49}$$

$$q_c = q_{icv} + q_{ir} \tag{4.50}$$

Table 4.22 Composition of the insulation layers

Module	1	2	3	4	5	6	7	8	9	10	11	12	13	14	15	16	17	18	19	20	21	22	23	24	25	26	27	28
Side Wall Layers (mm)																												
Cordierite brick	230	230	230	230	230	230	230	230	230	230	230	230	230	230	230	230	230	230	230	230	230	230	230	230	230	230	230	230
Refractory fiber	50	50	50	50	50	50	50	50	50	50	50	50	50	50	50	50	50	50	50	50	50	50	50	50	50	50	50	50
Refractory fiber	50	50	50	50	50	50	50	50	50	50	50	50	50	50	50	50	50	50	50									
Fiberglass	50	50	50	50	50	50	50	50	50	50	50	50	50	50	50	50	50	50	50									
Fiberglass										50	50	50	50	50	50	50	50	50	50									
Roof Layers (mm)																												
Cordierite plate	38	38	38	38	38	38	38	38	38											38	38	38	13	13	13	13	13	13
Cordierite brick	121	121	121	121	121	121	121	121	121	230	230	230	230	230	230	230	230	230	230	121	121	121						
Refractory concrete										10	10	10	10	10	10	10	10	10	10									
Refractory fiber	30	30	30	30	30	30	30	30	30	50	50	50	50	50	50	50	50	50	50	30	30	30	40	40	40	40	40	40
Fiberglass																							40	40	40	40	40	40

Table 4.23 Thermal conductivities of the different insulation materials

Thermal conductivity (W/m/K)											
°C	0	100	200	300	400	500	600	800	1000	1200	1400
Vermiculite grade A	0.0731	0.0831	0.0971	0.1131	0.1331	0.1622					
Vermiculite grade B	0.0641	0.0861	0.1101	0.1401	0.1812	0.2302					
Vermiculite grade C	0.0551	0.0881	0.1181	0.1551	0.2002	0.2623					
Refractory panel	0.0555	0.0727	0.0892	0.1059	0.1232	0.1421	0.1631	0.2146	0.2834	0.3750	0.4953
Refractory fiber	0.0125	0.0442	0.0743	0.1057	0.1404	0.1799	0.2246	0.3286	0.4427	0.5456	0.6039
Cordierite	0.3601	0.3693	0.3795	0.3901	0.4007	0.4109	0.4205	0.4376	0.4527	0.4686	0.4903

Table 4.24 Coefficients of the polynomial regressions for conductivities

Coefficients of polynomial regressions						
°C	C_0	C_1	C_2	C_3	C_4	C_5
Vermiculite grade A	0.0730691	6.2515E-05	5.0614E-07	−1.5165E-09	2.2497E-12	−6.32E-16
Vermiculite grade B	0.06406044	0.00020401	3.0424E-07	−2.11147E-09	7.705E-12	−7.3322E-15
Vermiculite grade C	0.05505268	0.00043525	−1.7725E-06	8.8834E-09	−1.7934E-11	1.418E-14
Refractory panel	0.05548612	0.00017766	−6.7967E-08	1.1819E-10		
Refractory fiber	0.01254138	0.00033498	−2.4859E-07	6.4785E-10	−3.0405E-13	
Cordierite	0.3601131	8.3926E-05	8.8703E-08	−1.3328E-10	5.3235E-14	

Table 4.25 Fiberglass thermal conductivity versus side temperatures

l (W/m/K)		Hot temperature (°C)							
W/m/K		0	100	200	300	400	500	600	700
Cold temperature (°C)	0	0.0348	0.0435	0.0534	0.0627	0.0766	0.0929	0.1138	0.1370
	100	0.0395	0.0476	0.0569	0.0673	0.0813	0.0999	0.1208	0.1446
	200	0.0464	0.0523	0.0627	0.0755	0.0906	0.1074	0.1300	0.1556
	300	0.0511	0.0581	0.0697	0.0824	0.0975	0.1200	0.1417	0.1660
	400	0.0557	0.0639	0.0766	0.0906	0.1074	0.1277	0.1521	0.1788
	500	0.0639	0.0720	0.0859	0.0999	0.1190	0.1417	0.1655	0.1951
	600	0.0755	0.0836	0.0975	0.1150	0.1324	0.1556	0.1823	0.2136
	700	0.0836	0.0952	0.1103	0.1277	0.1475	0.1718	0.1997	0.2322

$$q_c = k_1 \cdot S \cdot (T_1 - T_2) = \cdots = k_n \cdot S \cdot (T_{n-1} - T_n) \tag{4.51}$$

$$q_c = q_{ecv} + q_{er} \tag{4.52}$$

$$q_{ecv} = h_e \cdot S \cdot (T_n - T_0) \tag{4.53}$$

$$q_{er} = \sigma \cdot \varepsilon \cdot S \cdot \left(T_e^4 - T_0^4\right) \tag{4.54}$$

$$h = \frac{Nu \cdot k}{L} \tag{4.55}$$

where:

q_{icv} = internal convection heat transmission (kW).

h_i = internal convection coefficient heat transmission (kW/m^2/K).

S = surface of heat transmission (m^2).

T_i = temperature in the chamber of the kiln (K).

T_1 = temperature on the internal wall of the kiln (K).

q_{ir} = internal radiation heat transmission (kW).

σ = radiation constant (kW/m^2/K^4).

ε = internal radiation emissivity (dimensionless).

q_i = internal heat transmission (kW).

q_c = conduction heat transmission (kW).
k_i = conduction heat coefficient (kW/m^2/K).

Table 4.26 Polynomial coefficients of the fiberglass adjustment parameters. (*Source* Data compiled by the authors)

	a_0	a_1	a_2	a_3	a_4	a_5	a_6	a_7
C_1	0.03483	−0.0000117	9.56E-07	−4.59E-09	9.18E-12	−7.39E-15	1.75E-18	−5.59E-23
C_2	0.001956	1.88E-07	−2.06E-08	1.02E-10	−1.76E-13	4.87E-17	1.39E-19	−9.44E-23

T_i = temperature of the i insulation layer (K).

q_{ecv} = external convection heat transmission (kW).

q_{er} = external radiation heat transmission (kW).

h_e = external convection coefficient heat transmission (kW/m^2/K).

T_o = ambient temperature (K).

Nu = Nusselt's number (dimensionless).

L = characteristic dimension (m).

As the number of Nusselt and the conductivity of the gas are depending on the temperature of the fluid, the system has to be solved by means of an iterative process. Besides, the formula to calculate Nu varies as the geometric and thermodynamic conditions do. The interior of the kiln is modeled as a non-circular tube where the hydraulic diameter is calculated by:

$$D_h = 4 \cdot \frac{A_{cs}}{P_{\text{wet}}} \tag{4.56}$$

where:

D_h = hydraulic diameter (m).

A_{cs} = cross-sectional surface (m^2).

P_{wet} = wet perimeter (m).

A_{cs} and P_{wet} are shown in Fig. 4.29.

The thermodynamic characteristics of the flow are calculated for every temperature as follows:

$$\text{Re} = \frac{\rho \cdot v \cdot D_h}{\mu} \tag{4.57}$$

$$\text{Pr} = \frac{\mu \cdot C_p}{k_{\text{air}}} \tag{4.58}$$

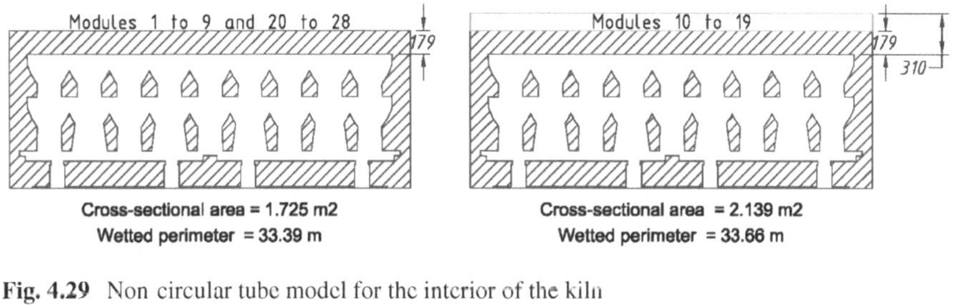

Fig. 4.29 Non circular tube model for the interior of the kiln

$$Gr = \frac{g \cdot \rho^2 \cdot L^3 \cdot \beta \cdot \Delta t}{\mu^2} \tag{4.59}$$

where:

Re = Reynold's number (dimensionless).

ρ = density (kg/m^3).

ν = velocity (m/s).

μ = dynamic viscosity (Pa/s).

Pr = Prandtl's number (dimensionless).

Cp = specific heat capacity (kJ/kg/K).

k_{air} = conductivity of the air (W/m/K).

Gr = Grashof's number (dimensionless).

β = expansion coefficient (K/m).

Dynamic viscosity of the air is calculated by the Shutherland's law:

$$\mu = \mu_0 \cdot \left(\frac{T}{T_0}\right) \cdot \frac{T_0 + 110.4}{T + 110.4}. \tag{4.60}$$

Density of the air is approximated by:

$$\rho = \rho_0 \cdot \frac{T_0}{T}. \tag{4.61}$$

Specific heat of the air is approximated by:

$$C_{p_{\text{air}}} = 0.9956 + 9.299 \cdot 10^{-5} T. \tag{4.62}$$

Thermal expansion coefficient is well approximated by:

$$\beta = \frac{1}{T}. \tag{4.63}$$

Inside the kiln, Nu is calculated according to Hausen's formula:

$$Nu = 3.66 + \frac{0.0668 \cdot \left(\frac{D_h}{L_{\text{mod}}} \cdot \text{Re} \cdot \text{Pr}\right)}{1 + 0.04 \cdot \left(\frac{D_h}{L_{\text{mod}}} \cdot \text{Re} \cdot \text{Pr}\right)^{\frac{2}{3}}}. \tag{4.64}$$

Outside the kiln, onto the walls, there is a natural vertical convection and the equation to obtain Nu is:

$$Nu = 0.555 \cdot (Gr \cdot \mathrm{Pr})^{\frac{1}{4}}. \tag{4.65}$$

Onto the roof, the convection natural and horizontal, thus the equation is:

$$Nu = 0.58 \cdot (Gr \cdot \mathrm{Pr})^{\frac{1}{5}}. \tag{4.66}$$

The flow characteristic in this chamber corresponds to a natural convection between horizontal plates, so the Nu in it is calculated by:

$$Nu = 0.068 \cdot Gr^{\frac{1}{3}}. \tag{4.67}$$

4.9.1.4 Load and Carts Thermal Evolution

Accordingly, with Fig. 4.25, the cart and the load thermal characteristics are shown in Table 4.27:

Equation 3.4 yields inlet and outlet enthalpies for every layer passing by each module at correspondent temperature.

Thermal power to heat up the supports is calculated based on the temperature difference inside the kiln between correspondent modules and specific heat of each material as shown below.

Table 4.27 Cart and load thermal characteristics

Cart masses	Material	m	Cp
		kg/cart	kJ/kg/K
Ware	Ceramic body	360	1.13
Supporting frames	Cordierite	112	1.46
Supporting plates	Cordierite	125	1.46
Supporting columns	Cordierite	18	1.46
Cart composition layers			
Layer-1	Refractory concrete	179	0.88
Layer-2	Thermal panel Kaowool	27	0.84
Layer-3	Thermal fiber Kaowool	25	1.10
Layer-4	Vermiculite concrete 1/8	1,229	0.24
Layer-5	Vermiculite concrete 1/5	222	0.24
Layer-6	Steel plate	92	0.46
Structure support and wheels	Iron	880	0.46

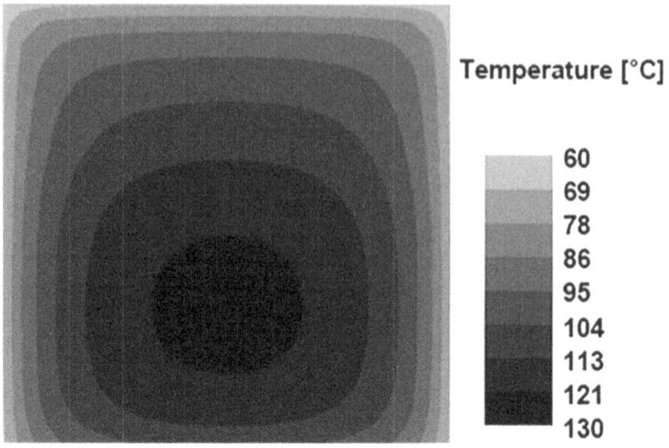

Fig. 4.30 Cross-section temperature gradient for a brick firing load at the outlet of the kiln

To compute the thermal load to heat up the cart, it has to be calculated the temperature gradient from inside chamber and floor below. So, temperature variability is different between the first layer and the wheels on bottom and supporting structure. This gradient states as well the conduction losses throughout the floor as calculated in Sect. 4.9.1.3.

Thermal power of the load along the kiln is calculated from the temperature on the surface of the load and the gradient into the ware. The temperature field is not uniform through the load and the layers of the cart. Nicolau et al. have established the gradient for a similar kiln for bricks manufacturing finding up to 50% of temperature change between surface and core of the load as Fig. 4.30 shows [8].

Equation (4.68) shows the way to compute the thermal requirement within every module in the tunnel. During the firing process, power is positive. On the contrary, during cooling process power is negative since the heat is transferred from masses to cooling air flows.

For every solid mass along the tunnel, it can be calculated the required power from mass flow, specific heat and temperature difference as:

$$\Delta H_i = \dot{m}_i \cdot C_i \cdot \Delta T_i. \tag{4.68}$$

where:

ΔH_i = enthalpy increase of the load (kW).

$\dot{m}_i$ = mass flow of the solid load (kg/s).

C_i = SPecific heat capacity (kJ/kg/K).

ΔT_i = temperature difference (K).

For masses over the level of the cart, the considered inlet and outlet temperatures correspond to the chamber of the kiln. However, for the different layers of the cart itself, the temperatures are defined by the gradient of the heat transfer toward the floor.

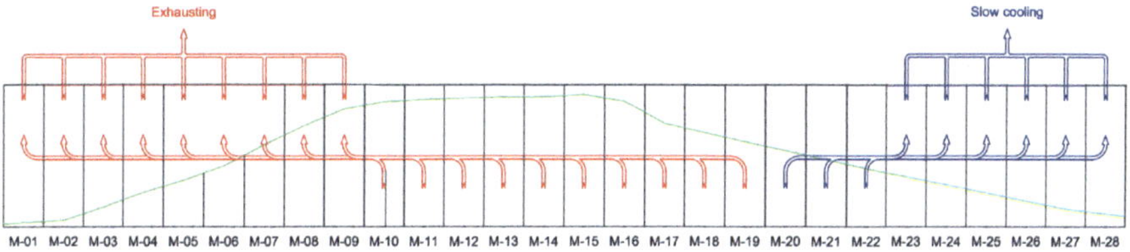

Fig. 4.31 Exhausting and cooling flows inside the kiln chamber

4.9.1.5 Mass and Energy Balances

Once calculated losses and enthalpies corresponding to solid masses, it only remains to compute gaseous flows by means of mass and energy balances as per (4.42) and (4.43) and Fig. 4.19 to each module as indicated in Fig. 4.26.

From resulting values, exhausting gases and air flows throughout the stacks are characterized to assess the possible recoverable heat. Figure 4.31 shows the configuration of the exhausting and cooling stacks.

In this figure, it can be appreciated how the exhausting fan produces the induced draught to bring heat upward.

4.9.2 Shuttle Kiln Model

The model of a shuttle kiln is quite similar to the tunnel one. The main difference lies in the way that cycle is carried out. The curve of temperature in the tunnel is spatial since it is completely present at any time in different sections of the tunnel. However, in a shuttle kiln, the whole camber is at the same temperature at every time, and the curve is temporal varying with time instead of location. Cross-sections of the kiln are shown in Fig. 4.32.

A general view of the shuttle kiln can be seen in Fig. 4.33 where different parts of the equipment are shown, so as the load inwards.

Specifications for a standard shuttle kiln of the leader of sanitary ware machinery manufacturer are as follows in Fig. 4.34.

The kiln capacity of the kiln chosen in this case study is 35 m^3. The characteristics of such a kiln are listed in Table 4.28.

The configuration of the carts is the same than the tunnel kiln one, although the dimensions differ each other.

To simulate or model the kiln, it has been conceived as a box enclosed by refractory blocks and the cart carrying the ware as Fig. 4.35 shows.

Figure 4.36 shows the temperature inside the chamber for the complete cycle.

As it happens to the tunnel kiln, there are different air flows into the kiln in order to help it to adjust the set curve:

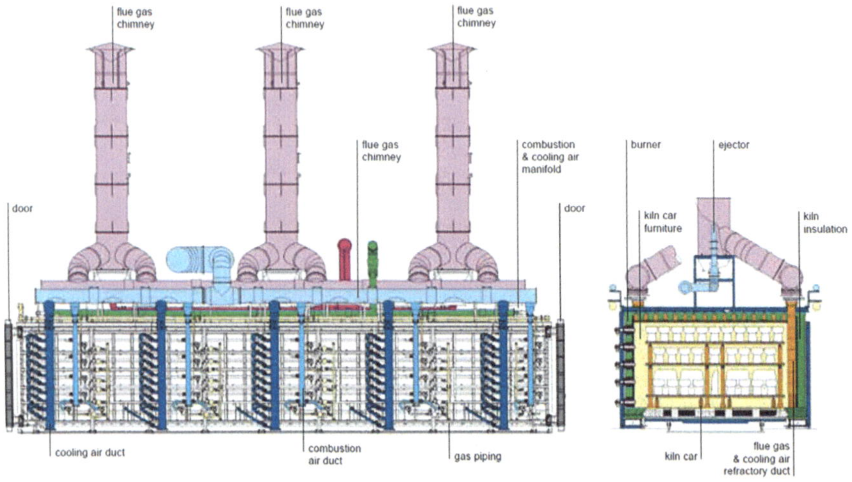

Fig. 4.32 Cross-section of a shuttle kiln (*Source* SACMI Catalogue)

Fig. 4.33 General view of a shuttle kiln (*Source* Data compiled by the authors)

- Combustion air to produce the oxidization of the NG as per the stoichiometric ratio.
- Excess of air is set 10% of the combustion air and allows complete combustion
- Injection air to be blown to rapidly adjust the excesses of temperature
- Diffusion/cooling air to provide uniformity of temperature through the kiln chamber during heating process and cooling the chamber after heating peak.

From this point on, the procedure to model the process is the same than the one for the tunnel with the result of the following model scheme (Fig. 4.37).

	HWS - FL
Possible kiln dimension Mögliche Ofengröße - Posibles dimensiones del horno	15 ÷ 305 m³
Useful width Nutzbreite - Anchura útil	4200 m ÷ 6400 mm
Possible car loading height Mögliche Ladehöhe d. Wagen - Altura de carga posible de la vagoneta	800 m ÷ 2450 mm
Useful car volume Nutzvolumen d. Wagen - Volúmen útil de la vagoneta	5.0 ÷ 23.5 m³
No. of cars in the kiln Wagenanzahl im Ofen - No. de vagonetas en el horno	3 ÷ 13
Average load per car (depending on product mix and loading conditions) Durchschn. Beladung pro Wagen (abhängig vom Produkt und Ladebedingungen) Carga media de la vagoneta (dependiente de la mezcla de productos y de las condiciones de carga)	6 pcs/m²
Maximum temperature Maximale Temperatur - Temperatura máxima	1300 °C
Firing temperature Brenntemperatur - Temperatura de cocción	up to 1260 °C
Approximate firing cycle Brennzyklus (ca.) - Ciclo de cocción aproximado	12 ÷ 30 h
Approximate specific net consumption (depending on technological and firing cycle conditions) Spezifischer Nettoverbrauch (ca.) (abhängig von verwendeter Technologie und Brennzyklus) Consumo neto específico aproximado (dependiente de las condiciones tecnológicas y del ciclo de cocción)	1550 ÷ 2200 kcal/kg
Specific gross consumption Spezifischer Bruttoverbrauch - Consumo bruto específico	750 ÷ 1070 kcal/kg

Fig. 4.34 Shuttle kiln technical specifications (*Source* SACMI Catalogue)

Table 4.28 Main characteristics of a shuttle kiln 35 m^3 (*Source* SACMI—Technical specifications)

Kiln specifications		
Kiln length	14,550	mm
Kiln width	6,200	mm
Kiln high	5,300	mm
Useful volume	35	m^3
Number of carts	9	
Temperature	1,300	°C
Sized power	5,760	kW
Duty cycle	14	h
Load rate	170	kg/m^3
Specific consumption	7,524	kJ/kg

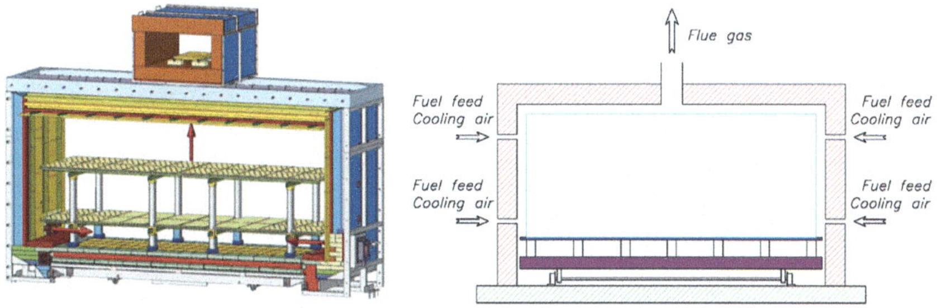

Fig. 4.35 Shuttle kiln cross-section (Courtesy of SACMI GROUP)

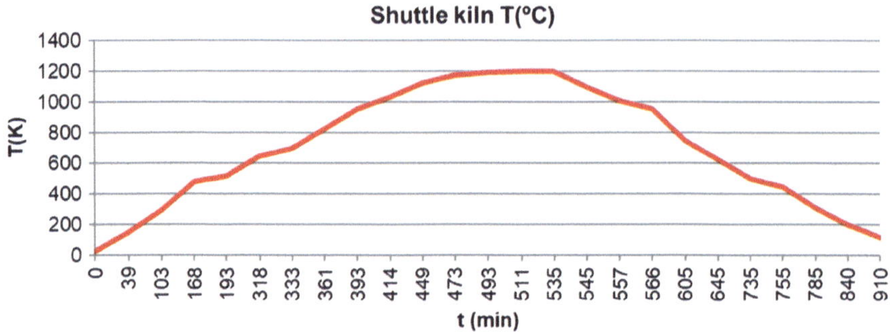

Fig. 4.36 Shuttle kiln temperature curve

Fig. 4.37 Thermal model of the shuttle kiln (*Source* Data compiled by the authors)

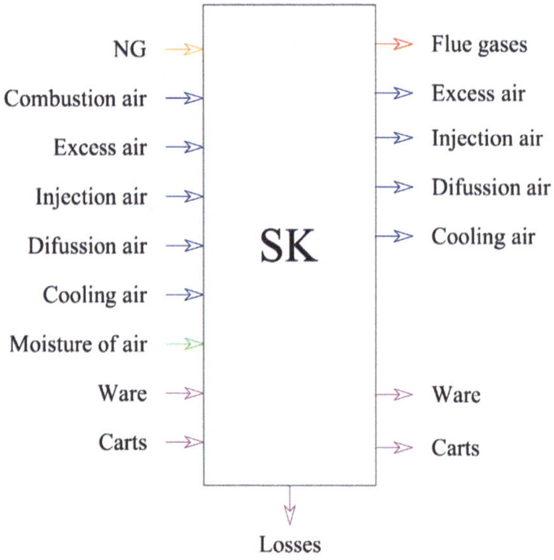

4.10 Sorting, Offices, Warehouse and Facilities Sections

The energy requirements for these sections are essentially electrical and pneumatic. Hence, as mentioned before, the equipment is characterized as generic electrical or pneumatic consumer as detailed in Sect. 3.2.1.

It is worth mentioning that there may be some boiler for showers. In such case, its thermal model has been already detailed in Sect. 4.5.2.1.

References

1. Ceramic technology sanitary-ware, 2010th edn. SACMI IMOLA S.C. (2010)
2. Manufacturing systems for ceramic sanitary ware SACMI. https://www.sacmi.com/en-US/Ceramics/Sanitaryware. Accessed 05 Nov 2020
3. Cuviella-Suárez C, Colmenar-Santos A, Borge-Diez D, López-Rey Á (2018) Management tool to optimize energy and water consumption in the sanitary-ware industry. J Clean Prod 197, Part 1:280–296. https://doi.org/10.1016/j.jclepro.2018.06.195
4. Cuviella-Suárez C, Colmenar-Santos A, Borge-Diez D, López-Rey Á (2019) Heat recovery in sanitary-ware industry applied to water and energy saving by multi-effect distillation. J Clean Prod 213:1322–1336. https://doi.org/10.1016/j.jclepro.2018.12.269
5. Saidur R, Hossain MS, Islam MR, Fayaz H, Mohammed HA (2011) A review on kiln system modeling. Renew Sustain Energy Rev 15(5):2487–2500. https://doi.org/10.1016/j.rser.2011.01.020
6. Reference document on best available techniques in the ceramic manufacturing industry. European Commission, August 2007 (2007). http://eippcb.jrc.ec.europa.eu/reference/BREF/cer_bref_0807.pdf. Accessed 02 Dec 2016
7. Kaya S, Küçükada K, Mançuhan E (2008) Model-based optimization of heat recovery in the cooling zone of a tunnel kiln. Appl Therm Eng 28(5):633–641. https://doi.org/10.1016/j.applthermaleng.2007.04.002
8. de Paulo Nicolau V, Dadam AP (2009) Numerical and experimental thermal analysis of a tunnel kiln used in ceramic production. J Braz Soc Mech Sci Eng 31(4):297–304. https://doi.org/10.1590/s1678-58782009000400003
9. Mezquita A, Boix J, Monfort E, Mallol G (2014) Energy saving in ceramic tile kilns: cooling gas heat recovery. Appl Therm Eng 65(1):102–110. https://doi.org/10.1016/j.applthermaleng.2014.01.002
10. Zhang Y, Wang J, Redemann T, Specht E (2015) Thermal behavior of kiln cars while traveling through a tunnel kiln. Adv Mech Eng 7(5):1687814015588468. https://doi.org/10.1177/1687814015588468

Chapter 5
Analysis of Consumptions

Once stated how to model every single process, this chapter tackles with the global model for the whole system along the manufacturing line. The procedure is

- to assess every magnitude to be evaluated through the manufacturing process
- to calculate the value of the representative variables; energy, power, masses, etc.
- to summarize the global sum of single processes into global demand
- to confront the obtained values for the two analyzed types of factory.

A database is produced based on the electrical equipment and its characteristics in order to split up the powers and energies through the different areas or processes of each technological production line as defined in Sect. 3.3 and Chap. 4.

5.1 Electrical Model

As stated in Sect. 3.2.1, the model the way to obtain the electrical assessment is to analyze all the electrical loads and their behavior throughout the duty cycle. The magnitudes to be characterized are:

- Power size and number of equipment units (kW)
- Duty cycle by means of "run-stop" hours
- Coefficients of use and simultaneity to get the averaged power
- Average power for every load (kW)
- The energy for one day and the whole year (MWh)
- Emissions of CO_2 from the electrical power generated in a fossil-fuel based power plant (t/year).

By means of the following equations applied to every facility within the process and their requirements, it can be computed the electrical consumption for a daily or annual cycle:

C. Cuviella-Suárez et al., *Water and Energy Use in Sanitary-ware Manufacturing*, Green Energy and Technology, https://doi.org/10.1007/978-3-030-72491-7_5

$$P_{\text{average}} = n_{\text{equipment}} \cdot P_{\text{sized}} \cdot f_u \tag{5.1}$$

where:

P_{average}: averaged electrical power

$n_{\text{equipment}}$: number of equipment units

P_{sized}: SIzed electrical power

f_u: coefficient of use.

$$W = P_{\text{average}} \cdot t_{\text{run}} \tag{5.2}$$

where:

W: energy for a defined period

t_{run}: time of operation within the defined period.

Provided compressed air requirement from these tables above, it can be computed the size of compressor needed as well as the electrical corresponding consumption by means of:

$$W_{\text{compressor}} = n_{\text{equipment}} \cdot Q_{\text{air}} \cdot f_u \cdot t_{\text{run}} \cdot W_{\text{specific}} \tag{5.3}$$

where W_{specific}: specific electrical energy to compress 1 Nm^3/h of air obtained from (3.1).

All the requirements for every section can be seen through the following tables, where sized powers, operation times and coefficients to adjust average versus sized power.

Regarding the lighting requirement, and following Sect. 3.2.1.2, Table 5.14 provides corresponding powers through the different sections of the factory.

5.2 Thermal Models

According to defined thermal models in Chap. 4, they are obtained thermal powers and energies for the different processes throughout the manufacturing line. Then, in order to sum them up, they are computed the total thermal requirement and residue. These values are crucial in order to optimize the energy resources. They state the thermal map within the factory to assess the possibilities of transferring heat between consumers in order to decrease the use of fuel. Following the procedure to build Tables 5.1, 5.2, 5.3, 5.4, 5.5, 5.6, 5.7, 5.8, 5.9, 5.10, 5.11, 5.12 and 5.13 for electrical demand, it can be deduced Table 5.15 for thermal requirements (Table 5.14).

Table 5.1 Electrical and pneumatic requirement for raw materials section

DB01	Raw materials	Number of equips	$P_{electric}$ (kW)	Q_{air} (Nm3/h)	RUN (h)	STOP (h)	Coef. use f_u
DB01-A	Mills	2	110	13	8	16	0.90
DB01-B	Belts	7	5		8	16	0.70
DB01-C	Pneumatic pumps	3	0	67	0	0	0.00
DB01-D	Hoppers	6	3	13	8	16	0.60
DB01-E	Lighting		3		8	16	0.86

Table 5.2 Electrical and pneumatic requirement for slip preparation section

DB02	Slip preparation	Number of equips	$P_{electrical}$ (kW)	Q_{air} (Nm^3/h)	RUN (h)	STOP (h)	Coef. use f_u
DB02-A	Blungers	3	125	7	8	16	0.60
DB02-B	Sieves	6	2	3	8	16	0.70
DB02-C	Peristaltic pumps	3	15	10	8	16	0.50
DB02-D	Slip tanks	7	15		24	0	1.00
DB02-E	Lighting		3		8	16	0.67

Table 5.3 Electrical and pneumatic requirement for glaze preparation section

DB03	Glaze preparation	Number of equips	$P_{electric}$ (kW)	Q_{air} (Nm^3/h)	RUN (h)	STOP (h)	Coef. use f_u
DB03-A	Mills	2	25	7	8	16	0.90
DB03-B	Belts	2	5	3	8	16	0.50
DB03-C	Blungers	2	30	3	8	16	0.60
DB03-D	Sieves	3	2	3	8	16	0.70
DB03-E	Pneumatic pumps	2	0	50	0	0	0.00
DB03-F	Glaze tanks	19	3	10	24	0	0.60
DB03-G	Lighting		4		8	16	0.68

5.2.1 Molds Dryer Model Results

According to the general procedure to assess the thermal model described in Sect. 4.4.1, the thermal balance of the model provides the thermal required power, which usually is supplied by natural gas. Table 5.16 shows the distribution of thermal consumption.

On the other hand, there is a residual source of heat from exhausting at 60 °C as shown in Table 5.17.

Table 5.4 Energy requirement for modeling and molds section

DB04	Modeling	Plaster mold factory							Modeling	Resin mold factory					
		Number of equips	$P_{electric}$	P_{gas}	Q_{air}	Run	Stop	Coef. use		Number of equips	$P_{electric}$	Q_{air}	Run	Stop	coef. use
			(kW)	(kW)	(Nm^3/h)	(h)	(h)	f_u			(kW)	(Nm^3/h)	(h)	(h)	f_u
DB04-A	Molds dryers	3	35	153	7	24	0	0.90	Resin mixer	1	5	7	8	16	0.90
DB04-B	Cranes	10	2		7	8	16	0.40	Cranes	1	2	7	8	16	0.40
DB04-C	Plaster preparation machine	1	5		17	8	16	0.90	Conditioning air	1	5	17	24	0	0.50
DB04-D	Pneumatic pumps	3	0		50	0	0	0.00	Pneumatic pumps	2	0	50	0	0	0.00
DB04-E	Lighting		18			8	16	0.83	Lighting		18		8	16	0.66

Table 5.5 Energy requirement for casting section

DB05	Plaster mold factory								Resin mold factory						
	Casting	Number of equips	$P_{electric}$	Q_{air}	Run	Stop	Coef. use	CASTING	Number of equips	$P_{electric}$	P_{gas}	Q_{air}	Run	Stop	Coef. use
			(kW)	(Nm³/h)	(h)	(h)	f_u			(kW)	(kW)	(Nm³/h)	(h)	(h)	f_u
DB05-A	Benches	30	20	33	16	8	0.60	Pressure casting machines	13	18		33	16	8	0.53
DB05-C	Ponds	2	5	7	8	16	0.60	Ponds	2	5		7	8	16	0.60
DB05-D	Tanks	2	25	10	24	0	1.00	Tanks	2	25		10	24	0	1.00
DB05-E	Pneumatic pumps	4	0	67	0	0	0.00	Pneumatic pumps	2	0		67	0	0	0.00
								Boilers	1	26	700		24	0	1.00
DB05-F	Lighting		41		8	16	0.63	Lighting		14			8	16	0.61

Table 5.6 Energy requirement for heating and ventilation section

DB06	Heating	Number of equips	$P_{electric}$	P_{gas}	Q_{air}	Run	Stop	Coef. use	Heating	Number of equips	$P_{electric}$	P_{gas}	Q_{air}	Run	Stop	Coef. use
			(kW)	(kW)	(Nm^3/h)	(h)	(h)	f_u			(kW)	(kW)	(Nm^3/h)	(h)	(h)	f_u
DB06-A	HVU	1	50	1500	7	24	0	0.50	HVU	1	50	200	7	24	0	0.50
DB06-B	Extraction fans	6	6			8	16	0.70	Extraction fans	6	6			8	16	0.70
DB06-C	Lighting		1			8	16	0.58	Lighting		1			8	16	0.58

Table 5.7 Energy requirement for drying section

DB07	Drying	Number of equips	$P_{electric}$	P_{gas}	Q_{air}	Run	Stop	Coef. use
			(kW)	(kW)	(Nm^3/h)	(h)	(h)	f_u
DB07-A	Ware dryers	2	55	695	7	24	0	0.90
DB07-B	Lighting		8			8	16	0.90

Table 5.8 Electrical requirement for glazing section

DB08	Glazing	Number of equips	$P_{electric}$	Q_{air}	Run	Stop	Coef. use
			(kW)	(Nm^3/h)	(h)	(h)	f_u
DB08-A	Carrousels	2	45	100	8	16	0.90
DB08-B	Glazing boots	8	5	50	8	16	0.70
DB08-C	Lighting		20		8	16	0.84

Table 5.9 Electrical requirement for kilns section

DB09	Kilns	Number of equips	$P_{electric}$	P_{gas}	Q_{air}	Run	Stop	Coef. use
			(kW)	(kW)	(Nm^3/h)	(h)	(h)	f_u
DB09-A	Tunnel kiln	1	125	2,017	33	24	0	0.60
DB09-B	Shuttle kiln	1	100	2,500	17	20	4	0.60
DB09-C	Second fire boot	2	5		7	8	16	0.70
DB09-D	Lighting		10			8	16	0.60

Table 5.10 Energy requirement for sorting section

DB10	Sorting	Number of equips	$P_{electric}$	P_{gas}	Q_{air}	Run	Stop	Coef. use
			(kW)	(kW)	(Nm^3/h)	(h)	(h)	f_u
DB10-A	Sorting lines	4	5		23	8	16	0.90
DB10-B	Wrapping machine	1	10	400	3	8	16	0.70
DB10-C	Lifting tables	4	5		10	8	16	0.50
DB10-D	Lighting		8			8	16	0.70

5.2.2 HVU Model Results

As explained in Sect. 4.6.3, HVU depends strongly on the ambient condition. Temperature, humidity and solar radiation define the thermal energy required to keep the hygrometric conditions in the casting hall.

Table 5.11 Energy requirement for offices section

DB11	Offices	Number of equips	P_{electric} (kW)	Run (h)	Stop (h)	Coef. use f_u
DB11-A	Office desk	15	1	8	16	0.70
DB11-B	IT central servers	1	10	24	0	0.60
DB11-C	Conditioning air	10	5	8	16	0.40
DB11-D	Other consumers	1	20	8	16	0.70
DB11-E	Lighting		4	8	16	0.54

Table 5.12 Energy requirement for ware-house

DB12	Ware house	Number of equips	P_{electric} (kW)	RUN (h)	STOP (h)	Coef. use f_u
DB12-A	Office	1	3	8	16	0.70
DB12-B	Batteries charging	5	1	24	0	0.90
DB12-C	Lighting		4	8	16	0.83

Figures 5.1 and 5.2 and show the clear difference between casting (8–16 h) and drying periods in a plaster mold-based factory due to the amount of water to be evaporated during night time. On the other hand, for a resin factory, it can be seen how that difference is notably lower though there is a valley on the curves due to climate influence. Heat demand is lower because there is no water to be evaporated from molds.

It can be seen in this figure the impact of climate condition. At noon, when radiation and temperature are higher, the thermal requirement is lower. On the other hand, maximum demand occurs in winter time. Of course, minimum corresponds to summer warmest weather.

For the purpose of this case study, the annual averaged curve of demand has been taken to assess the thermal model. Knowing the environmental parameters of casting hall, a proper automatized system would be able to optimize the energy requirements [1] (Fig. 5.3).

Under such conditions, the averaged thermal balance for the whole daily cycle is shown in Table 5.18.

It can be seen how during day time solar irradiation produces negative losses. It means that, actually, heat transfer direction is led inwards.

The averaged powers split up for both types of factories can be seen. The requirement for casting period where only water from ware is evaporated is relatively low for both factories. It is lower for the plaster mold because this period coincides with solar irradiation. On the contraire, during drying period, where is evaporated all the water contained in the molds, it is required a much greater thermal power because of the following factors:

Table 5.13 Energy requirement for facilities

DB13	Facilities	Number of equips	$P_{electric}$ (kW)	P_{gas} (kW)	Q_{air} (Nm^3/h)	RUN (h)	STOP (h)	Coef. use f_u
DB13-A	Compressors	2	160			24	0	0.40
DB13-B	Boilers	1	5	110		24	0	0.60
DB13-C	Maintenance work shop	1	45		17	8	16	0.50
DB13-D	Losses hopper and crushing machine	1	50		13	8	16	0.50
DB13-E	Pumps and reverse osmosis plant	10	5		3	24	0	0.60
DB13-F	Wastewater plant	1	75		10	24	0	0.40
DB13-G	Security room	1	5			24	0	0.50
DB13-H	Waste treatment	1	5		7	8	16	0.40
DB13-I	Lunch room	1	5			8	16	0.30
DB13-J	Canteen	1	10	20		8	16	0.30
DB13-K	External lighting	1	10			12	12	1.00
DB13-L	Plugs circuit	1	32			24	0	0.70
DB13-M	Watering	1	0			0	0	0.00
DB13-N	Electrical station	1	2			24	0	0.30
DB13-O	Gas station	1	1			24	0	0.30
DB13-P	Generator	1	1			24	0	0.30
DB13-Q	Water tank reservoir	1	0			0	0	0.00
DB13-R	Lighting		9			8	16	0.46

- Contained water in the mold triples the water in the ware
- The temperature of the hall for drying process is 15 °C greater
- During drying period, there is not solar radiation
- External temperature during drying is lower and so the thermal losses.

It is important to remark the thermal solar influence during midday. The thermal fuel requirement drops to zero. This happens because the solar radiation provides all the thermal power to supply thermal losses as well as evaporations and heating of flows. Besides, there is a surplus of energy that produces a temperature increase in the casting hall, building, etc.

Table 5.14 Lighting levels and corresponding powers. (*Source* Data compiled by the authors)

Lighting levels	Lighting type	Efficiency	Plaster mold factory				Resin mold factory			
			%	*S*	*E*	$P_{conventional}$	%	*S*	*E*	$P_{convencional}$
		lm/W		m²	lx	kW		m²	lx	kW
Raw Materials	Vapor Na AP	100	4.15%	1,350	200	3.38	14.32%	1,350	200	3.38
Slip preparation	Vapor Na AP	100	3.07%	1,000	200	2.50	10.61%	1,000	200	2.50
Glaze preparation	Fluorescence	80	2.31%	750	300	3.52	7.96%	750	300	3.52
Modeling and Molds	Fluorescence	80	6.92%	2,250	500	17.58	23.87%	2,250	500	17.58
Casting	Fluorescence	80	19.98%	6,500	400	40.63	22.99%	2,167	400	13.54
Heating and Ventilation	Fluorescence	80	0.77%	250	200	0.78	2.65%	250	200	0.78
Drying	Fluorescence	80	7.69%	2,500	200	7.81	26.53%	2,500	200	7.81
Glazing	Fluorescence	80	7.69%	2,500	500	19.53	26.53%	2,500	500	19.53
Kilns	Fluorescence	80	9.99%	3,250	200	10.16	34.48%	3,250	200	10.16
Sorting	Fluorescence	80	3.07%	1,000	500	7.81	10.61%	1,000	500	7.81
Offices	Fluorescence	80	1.54%	500	500	3.91	5.31%	500	500	3.91
Ware house	Fluorescence	80	3.84%	1,250	200	3.91	13.26%	1,250	200	3.91
Facilities*	Various	98	28.98%	9,425	98	9.02	103.71%	9,775	98	9.83
Total Lighting=			100.00%	32,525		130.52	100.00%	28,542		104.24
*Facilities Split up										
Compressors	Fluorescence	80	0.31%	100	200	0.31	1.06%	100	200	0.31
Boilers	Fluorescence	80	0.08%	25	200	0.08	0.27%	25	200	0.08
Maintenance work shop	Fluorescence	80	0.77%	250	300	1.17	2.65%	250	300	1.17
Losses hopper and crushing machine	Vapor Na AP	100	1.54%	500	100	0.63	5.31%	500	100	0.63

(continued)

Table 5.14 (continued)

Lighting levels	Lighting type	Efficiency	Plaster mold factory				Resin mold factory			
			%	S	E	$P_{conventional}$	%	S	E	$P_{convencional}$
		lm/W		m^2	lx	kW		m^2	lx	kW
Pumps and reverse osmosis plant	Fluorescence	80	0.31%	100	200	0.31	3.18%	300	200	0.94
Wastewater plant	Vapor Na AP	100	1.08%	350	100	0.44	5.31%	500	100	0.63
Security room	Fluorescence	80	0.06%	20	300	0.09	0.21%	20	300	0.09
Waste treatment	Fluorescence	80	0.03%	10	100	0.02	0.11%	10	100	0.02
Lunch room	Fluorescence	80	0.15%	50	200	0.16	0.53%	50	200	0.16
Canteen	Fluorescence	80	0.06%	20	200	0.06	0.21%	20	200	0.06
External lighting	Vapor Na AP	100	23.06%	7,500	50	4.69	79.58%	7,500	50	4.69
Plugs circuit							0.00%			
Watering							0.00%			
Electrical station	Fluorescence	80	0.46%	150	200	0.47	1.59%	150	200	0.47
Gas station	Vapor Na AP	100	0.15%	50	200	0.13	0.53%	50	200	0.13
Generator	Fluorescence	80	0.15%	50	200	0.16	0.53%	50	200	0.16
Water tank reservoir	Vapor Na AP	100	0.77%	250	100	0.31	2.65%	250	100	0.31

Table 5.15 Thermal demand by consumer

Thermal demand calculation												
Facilities	Plaster mold factory						Resin mold factory					
	Q	RUN	STOP	Coef. use	$P_{average}$	W_{year}	Q	RUN	STOP	Coef. use	P_{media}	W_{year}
	(kW)	(h)	(h)	f_u	(kW)	(kWh/year)	(kW)	(h)	(h)	f_u	(kW)	(kWh/year)
Molds dryers	458	24	0	0.79	360	2,004,792						
Boilers							700	8	16	0.86	603	1,118,246
Heating and ventilation unit	1,500	24	0	0.52	782	4,351,851	200	24	0	0.55	110	609,956
Ware dryers	695	22	2	0.60	414	2,112,524	695	22	2	0.60	414	2,112,525
Tunnel kiln	2,017	24	0	1.00	2,017	14,524,898	2,017	24	0	1.00	2,017	14,524,898
Shuttle kiln	2,500	15	9	0.38	940	4,275,871	2,500	15	9	0.38	940	4,275,871
Wrapping machine	400	8	16	0.25	100	240,000	400	8	16	0.25	100	240,000
Boilers	110	24	0	0.95	105	581,856	110	24	0	0.95	105	581,856
Canteen	20	8	16	0.50	10	18,560	20	8	16	0.50	10	18,560
Total gas=	7,701	24	0	0.61	4,727	28,110,351	6,643	24	0	0.65	4,298	23,481,912

Table 5.16 Thermal requirement of a mold dryer

MD thermal balance		
	kW	%
$Q_{supports}$	0.25	0.16
Q_{molds}	0.85	0.56
$Q_{vaporization}$	100.67	65.91
Q_{air}	13.68	8.96
$Q_{exhausting}$	20.66	13.52
Q_{losses}	16.64	10.89
Total	152.75	100

Table 5.17 Residual heat for the mold dryer

Exhausting	m_{OUT}	H_{OUT}	T
	kg	kJ	K
CO_2	5,293	269,507	333
H_2O	4,330	11,314,564	333
Excess air	3,666	221,070	333
N_2	28,961	1,806,787	333
Mixture	42,250	13,611,927	333

Power variation along the averaged cycle is as shown in Fig. 5.4.

Exhausting flow does not matter from the recovering point of view, since the temperature is too low and it is very humid (18% for casting period and 28% for drying). Table 5.19 shows composition.

This exhausting flow is composed by:

- Air required to absorb water from ware and the moisture that it contains
- Air required to absorb water from molds and the moisture that it contains (only for drying period)
- Products of combustion
- Moisture contained in the combustion air
- Water evaporated from ware
- Water evaporated from molds (only for drying period).

5.2.3 *Ware Dryer Model Results*

According to Sect. 4.7.1, there are four heating stages and one cooling stage up to hall temperature, where the heat flow is outward from the hot masses inside the chamber. The results of the needed heat to operate the ware dryer by the energy and mass balances are shown in Table 5.20.

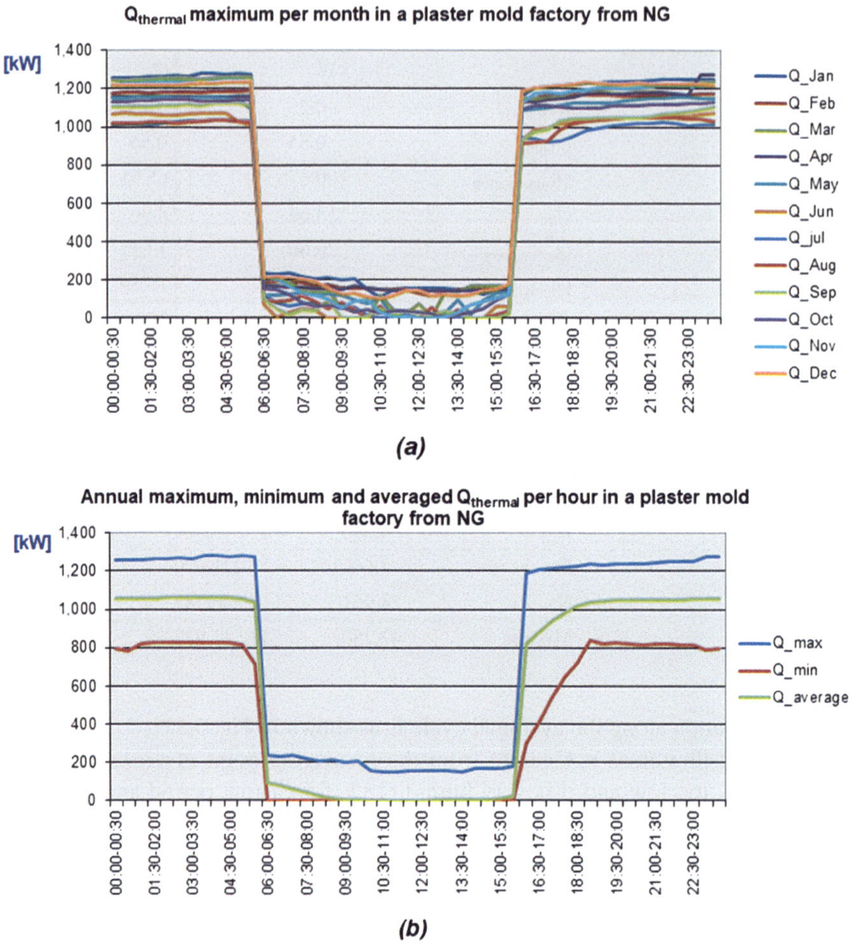

Fig. 5.1 Thermal power required for HVU in a plaster mold factory. **a** Averaged power every month. **b** Max, min and annual averaged values

Table 5.21 splits up the distribution of thermal power required in the dryer in order to heat up every flow as per Sect. 4.7.2.

As well as it happens in the HVU's, the exhausting is rather cool and humid. Thus, there is not much interest into recover this heat.

Figure 5.5 shows the variability of the thermal demand for the ware dryer (Table 5.22).

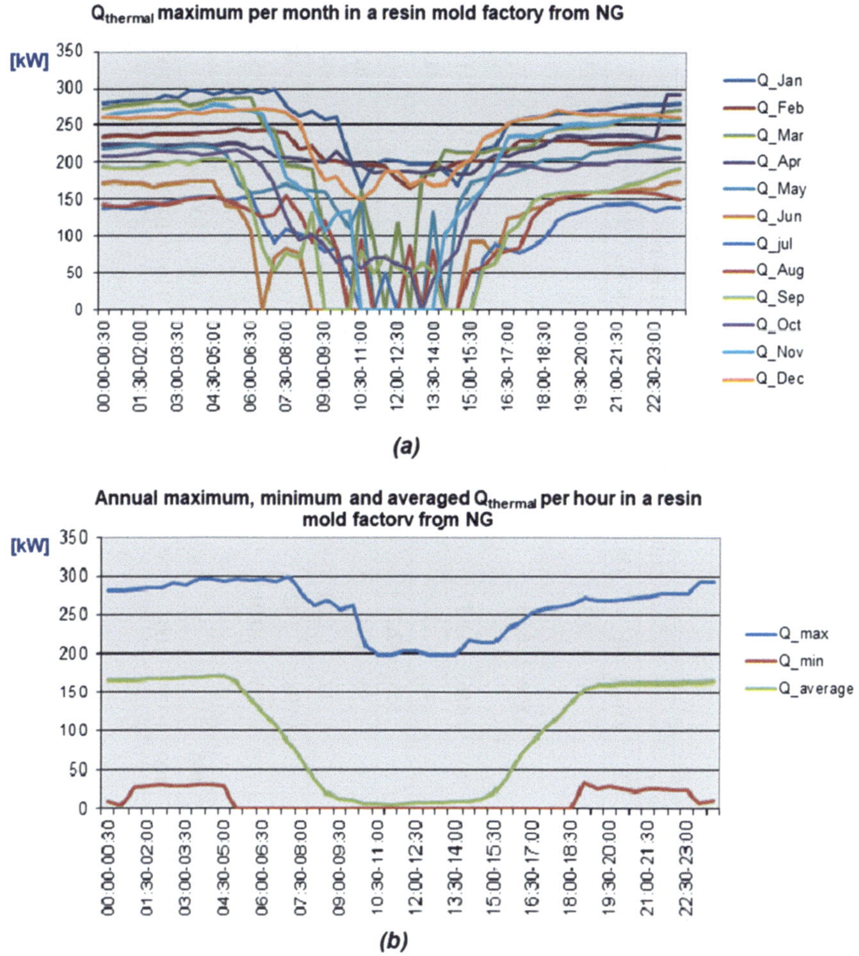

Fig. 5.2 Thermal power required for HVU in a resin mold factory. **a** Averaged power every month. **b** Max, min and annual averaged values

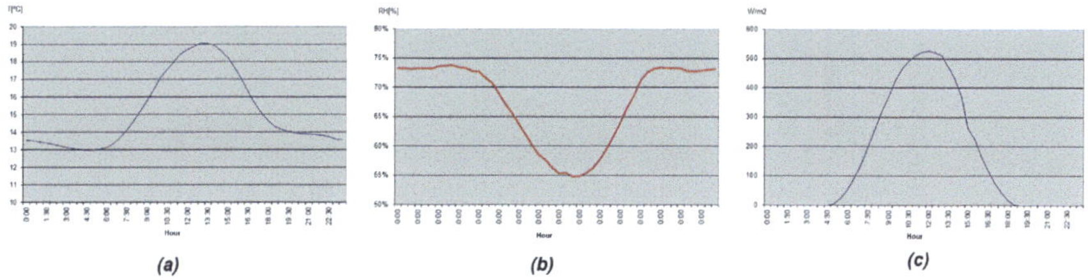

Fig. 5.3 Averaged curve of climate conditions (*Source* Data compiled by the authors)

Table 5.18 Thermal requirements of a HVU

Thermal requirements for heating and ventilation	Plaster mold factory												Resin mold factory			
	Casting				Drying				Daily				Casting			
	kW_{in}	%	kW_{out}	%	kW_{in}	%	kW_{out}	%	kW_{in}	%	kW_{out}	%	kW_{in}	%	kW_{out}	%
Fresh air to evaporate from ware			21	91.29			31	2.32			33	4.24			26	23.32
Fresh air to evaporate from mold							356	26.93			148	18.97			–	–
Combustion reactants			3	11.75			156	11.77			92	11.77			13	11.75
Vaporizing ware water			49	207.63			48	3.63			48	6.18			49	44.49
Vaporizing mold water							558	42.21			328	41.93			–	–
Losses side walls			18	78.48			59	4.45			42	5.38			22	20.31
Losses floor			9	39.55			17	1.28			14	1.76			9	8.47
Losses roof			−301	−1282.71			98	7.42			−68	−8.73			−102	−93.47
$T_{increase}$ of building, hall environment,…			224	954.01			0	−0.01			145	18.49			93	85.11
NG heating value	23	100			1,323	100			782	100			110	100%	–	
Total heat requirement	23	100	23	100	1,323	100	1,323	100	782	100	782	100	110	100%	110	100

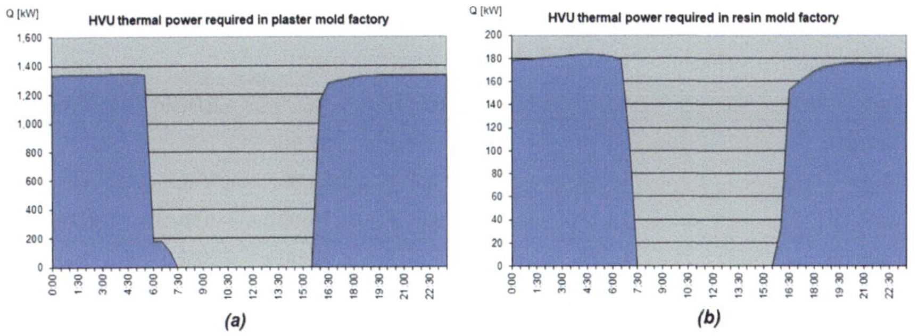

Fig. 5.4 Averaged thermal demand for the HVU in a daily cycle

Table 5.19 Exhausting composition of HVU's

Exhausting	PMF						RMF		
	Casting			Drying			Casting		
	m_{out}	H_{out}	T	m_{out}	H_{out}	T	m_{out}	H_{out}	T
	kg/cycle	kJ/cycle	°C	kg/cycle	kJ/cycle	°C	kg/cycle	kJ/cycle	°C
Fresh drying ware air	93,072	2,336,543	25	58,271	2,341,392	40	220,511	5,535,889	25
Fresh drying ware moist	684	1,741,319	25	418	1,075,753	40	1,597	4,069,401	25
Fresh drying mold air	0	0	25	677,303	27,214,979	40			25
Fresh drying mold moist	0	0	25	4,855	12,503,924	40			25
CO_2	18	1,005	25	1,389	128,126	40	542	11,259	25
H_2O	70	100,915	25	5,557	8,050,242	40	444	1,130,202	25
N_2	239	6,219	25	18,882	785,178	40	2,680	69,646	25
Combustion air moist	2	5,430	25	175	450,804	40	25	62,955	25
Ware vaporized water	718	1,830,049	25	1,006	2,590,224	40	1,724	4,392,117	25
Mold vaporized water	0	0	25	11,690	30,107,254	40			
Total	94,803	6,021,479	25	779,545	85,247,876	40	227,524	15,271,468	25

Table 5.20 Thermal operation parameters for a ware dryer

Drying stage	Time	Evaporation	Air to evaporate			Q
	h	kg	kg	°C	RH(%)	kW
Heating 1	2	539	26,061	37.50	75%	357
Heating 2	3	754	23,503	50.00	50%	336
Heating 3	2	1,078	10,352	75.00	30%	695
Heating 4	3	1,293	4,717	90.00	20%	480
Cooling 5	1	–	–	–	–	–
Average		852	14,317	64.50	42%	414

Table 5.21 Thermal power breakdown for the ware dryer

Heating requirements	Drying stage					
	1	2	3	4	5	Average
	kW	kW	kW	kW	kW	kW
Fresh air heating	45	55	72	29	0	52
Fresh air moisture heating	1	1	1	1	0	1
Exhausting	44	43	97	70	0	54
Ware heating	31	21	63	25	0	0
Carts heating	12	8	23	9	0	0
Cooling air heating	0	0	0	0	– 415	37
Water vaporization	212	185	392	286	0	234
Losses	12	24	47	61	31	37
Total	357	336	695	480	– 385	414

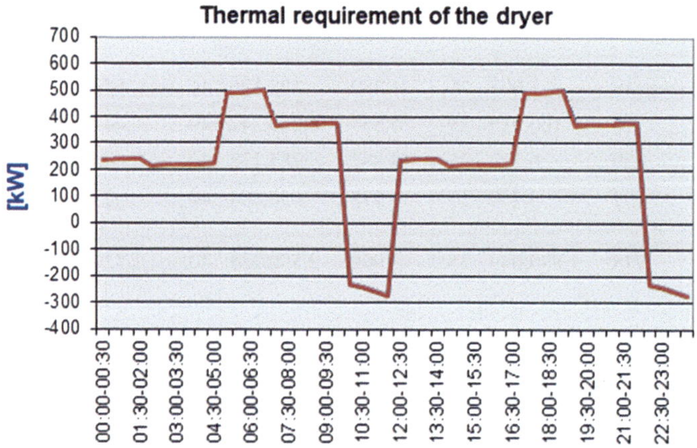

Fig. 5.5 Thermal demand of a ware dryer for the whole daily cycle

Table 5.22 Exhausting composition of a ware dryer. (*Source* Data compiled by the authors)

Exhausting	Drying stage									
	1		2		3		4		5	
	kg	kJ	kg	kJ	kg	kJ	kg	kJ	kg	kJ
Air for drying	26,061	981,643	23,503	1,180,780	10,352	780,673	4,717	427,143		
Air moisture	288	740,049	260	673,487	114	301,992	52	139,095		
CO_2	147	4,619	208	8,764	287	18,421	297	23,101		
H_2O	120	309,459	170	440,998	235	619,879	243	649,179		
N_2	727	28,347	1,027	53,383	1,418	110,587	1,469	137,537		
Combustion air moisture	10	26,726	15	38,086	20	53,535	21	56,066		
Cooling air									45,750	2,643,783
Vaporized water	539	1,384,993	754	1,956,616	1,078	2,845,688	1,293	3,451,344		
TOTAL	27,892	3,475,835	25,936	4,352,113	13,503	4,730,775	8,093	4,833,465	45,750	2,643,783

5.2.4 *Tunnel Kiln Model Results*

5.2.4.1 Energy Balance for Each Module

As detailed in Sect. 4.9.1, the result of thermal requirements for every module of the kiln is shown in Table 5.23.

The columns in Table 5.23 refer to:

- Heat transfer throughout enclosures (side walls, roof and floor).
- Heat transfer to the load (ware, cart and supports). For the heating section, it is positive to heat it up. For the cooling section, the accumulated heat is delivered by means of the cooling
- $Q_{\text{combustion}}$ represents the balance of combustion as the difference of exhausting gases and combustion reactants. It is negative because exhausting energy is lower than fuel plus reactants enthalpies.
- $Q_{\text{roof air}}$ represents the energy required to heat the roof air up.
- Q_{induced} draught represents the air flow draught from module downwards. Except for modules 15 and 16, it is negative due to temperature gradient. The air comes hotter from module downwards. This is a key for efficiency. The more draught induced, the less fuel required to heat up into the module.
- $Q_{\text{rapid cooling}}$ represents the air into modules 17–19 in order to produce the rapid cooling required for ceramic sintered properties. This air is sucked upward to exhausting stack, so it has to be heated up.
- $Q_{\text{slow cooling}}$ represents the air into modules 20–28 in order the final cooling of the ware. It takes the heat from load and is delivered throughout the slow cooling stack.

5.2.4.2 Global Energy Balance

Table 5.23 provides values referred to the individual balance for each module. When combined, it can be obtained the global balance for the whole kiln as shown in Table 5.24.

The total sum of inlet and outlet enthalpies includes chemical and physical values of enthalpy referred to normal conditions. On the contrary, the Q values of energy represent how the heat value of the fuel is distributed for each thermal requirement within the kiln.

5.2.4.3 Residual Heat Throughout the Stacks

The total available power to be recovered from kiln is provided by the exhausting stacks. The thermodynamic important characteristics of the air-gas flows through them and the mixture are shown in Table 5.25.

Potential heat recovery can be summarized as stated in Table 5.26.

Table 5.23 Thermal requirements in a tunnel kiln

Zone	Module	X	T	Q_{walls}	Q_{roof}	Q_{floor}	Q_{losses}	Q_{load}	$Q_{combustion}$	$Q_{roof\ air}$	$Q_{induced\ draught}$	$Q_{rapid\ cooling}$	$Q_{slow\ cooling}$
		m	°C	kW	kW	kW	kW	kW	kW	kW	kW	kW	kW
Preferring	1	1.5	43	0.01	0.03	0.02	0.07	40.72	−23.33	0.63	−18.09		
	2	4.5	62	0.03	0.07	0.03	0.13	111.69	0.00	1.29	−124.16		
	3	7.5	183	0.18	0.40	0.14	0.72	210.73	−72.06	5.54	−144.93		
	4	10.5	317	0.54	1.04	0.32	1.90	205.22	−93.01	10.35	−124.47		
	5	13.5	427	1.06	1.82	0.50	3.38	196.76	−54.87	14.40	−159.67		
	6	16.5	559	2.01	3.11	0.75	5.86	260.22	−27.67	19.39	−257.80		
	7	19.5	754	4.22	5.66	1.17	11.04	287.22	−88.24	27.00	−237.02		
	8	22.5	926	7.07	8.49	1.59	17.14	251.83	−84.52	33.92	−218.38		
	9	25.5	1,081	10.41	11.43	2.00	23.84	167.06	−148.33	40.29	−82.85		
Firing	10	28.5	1,145	10.90	10.57	2.18	23.65	65.18	−62.52		−26.31		
	11	31.5	1,167	11.43	10.86	2.24	24.54	25.70	−36.63		−13.61		
	12	34.5	1,179	11.73	11.11	2.27	25.12	15.86	−31.26		−9.72		
	13	37.5	1,188	11.95	11.30	2.30	25.56	11.32	−30.73		−6.15		
	14	40.5	1,194	12.10	11.43	2.32	25.85	19.60	−25.89		−19.56		
	15	43.5	1,214	12.61	11.86	2.38	26.85	−30.97	−49.62		53.74		
	16	46.5	1,153	11.09	10.58	2.20	23.87	−203.02	0.00		179.15		
Rapid cooling	17	49.5	947	6.76	6.79	1.64	15.19	−256.97				241.78	
	18	52.5	818	4.66	4.86	1.32	10.83	−201.18				190.35	
	19	55.5	688	3.01	3.29	1.02	7.32	−141.67				134.35	
Slow cooling	20	58.5	637	3.58	4.03	0.91	8.52	−81.20					72.68

(continued)

Table 5.23 (continued)

Zone	Module	X	T	Q_{walls}	Q_{roof}	Q_{floor}	Q_{losses}	Q_{load}	$Q_{combustion}$	$Q_{roof\ air}$	$Q_{induced\ draught}$	$Q_{rapid\ cooling}$	$Q_{slow\ cooling}$
		m	°C	kW	kW	kW	kW	kW	kW	kW	kW	kW	kW
	21	61.5	585	2.92	3.40	0.80	7.12	−81.68					74.56
	22	64.5	534	2.34	2.83	0.70	5.87	−100.36					94.49
	23	67.5	460	1.64	3.17	0.56	5.37	−119.47					114.10
	24	70.5	386	1.10	2.26	0.43	3.79	−121.05					117.26
	25	73.5	313	0.69	1.52	0.32	2.52	−123.30					120.78
	26	76.5	239	0.39	0.94	0.21	1.54	−127.34					125.80
	27	79.5	165	0.19	0.51	0.12	0.82	−95.22					94.40
	28	82.5	133	0.13	0.36	0.08	0.57	−35.77					35.20

Table 5.24 Balance for a tunnel kiln

In	H_{in}	m_{in}	Out	H_{out}	m_{out}	DH	Q_{fuel}	$Q_{required}$	%in	%out
	kW	kg/s		kW	kg/s	kW	kWin	kWout		
Ware	11	0.4038	Ware IN	46	0.4038	Q_{load}		34		1.70%
Plates	5	0.1401	Plates IN	20	0.1401	$Q_{supports}$		31		1.55%
Columns	1	0.0199	Columns IN	3	0.0199					
Supports	5	0.1260	Supports IN	18	0.1260					
Refractory concrete cart	4	0.2013	Refractory concrete cart	18	0.2013	Q_{cart}		84		4.18%
Refractory panel cart	1	0.0298	Refractory panel cart	2	0.0298					
Refractory fiber cart	1	0.0280	Refractory fiber cart	3	0.0280					
Vermiculite 1/8 cart	8	1.3781	Vermiculite 1/8 cart	33	1.3781					
Vermiculite 1/5 cart	1	0.2494	Vermiculite 1/5 cart	6	0.2494					
Steel plate cart	1	0.1030	Steel plate cart	5	0.1030					
Wheels cart	11	0.9872	Wheels cart	45	0.9872					
CH_4 heat value	2,017	0.0420				Q_{NG}	2,017		100.00%	
CH_4 Physical enthalpy	2					$Q_{exhausting}$		716		35.52%
O_2 combustion air	4	0.1681	CO_2	29	0.1156					
N_2 combustion air	14	0.5534	H_2O from combustion	284	0.0946					
O_2 excess air	0	0.0168	Excess air	19	0.0721					
N_2 excess air	1	0.0553	N_2	153	0.5534					
Roof air	8	0.3120	Roof air	84	0.3120					
Induced draught	2	0.0749	Induced draught	20	0.0749					
Rapid cooling	16	0.6553	Rapid cooling	176	0.6553					

(continued)

Table 5.24 (continued)

In	H_{in}	m_{in}	Out	H_{out}	m_{out}	DH	Q_{fuel}	$Q_{required}$	%in	%out
	kW	kg/s		kW	kg/s	kW	kWin	kWout		
Slow cooling	75	2.9677	Slow cooling	916	2.9677	$Q_{slow\ cooling}$		842		41.74%
			Side walls losses	135		$Q_{side\ walls}$		135		6.68%
			Roof losses	144		Q_{roof}		144		7.12%
			Floor losses	30		Q_{floor}		30		1.51%
Total In	2,190	8.5121	Total Out	2,190	8.5121	Balance	2,017	2,017	100.00%	100.00%

Table 5.25 Flue gases and air from a tunnel kiln

Exhausting per stack				
Stack	Fluid	m	T	Q
		kg/s	°C	kW
Exhausting	CO_2	0.1156	264	29
	H_2O	0.0946	264	284
	Air excess	0.0721	264	19
	N_2	0.5534	264	153
	Roof air	0.3120	264	84
	Induced draught	0.0749	264	20
	Rapid cooling	0.6553	264	176
Slow cooling	Slow cooling	2.9677	303	916
Residual heat		4.8455	288	1,681

Table 5.26 Recoverable heat

Residual heat	°C	kg/s	kW
Exhausting stack	264	1.8779	765
Slow cooling stack	303	2.9677	916
Total=	288	4.8455	1,681

1,681 kW from residual heat means 77% of the total inlet enthalpy at a reasonable level of temperature to be recovered.

5.2.5 *Shuttle Kiln Model Results*

5.2.5.1 Energy Balance at Each Stage

As detailed in Sect. 4.9.2, the result of thermal requirements for every stage of kiln cycle n is shown in Table 5.27.

The columns in this Table refer to:

- Heat transfer to the ware and the rest of the load (supports and cart). For the heating cycle, it is positive to heat it up. For the cooling cycle, the accumulated heat is returned.
- Heat is required for combustion purposes. It results negative because it has been calculated as the difference of combustion products minus reactants. This is the energy supplied by the fuel to provide the thermal requirement.
- Heat is required to increase the temperature of different blown air into the chamber (cooing, diffusion and injection).

Table 5.27 Thermal requirements in a shuttle kiln

	Stage	$T_{average}$	Q_{ware}	Q_{load}	$Q_{combustion}$	$Q_{cooling\ air}$	$Q_{difussion\ air}$	$Q_{injection\ air}$	$Q_{side\ walls}$	Q_{roof}	Q_{floor}
		°C	kJ	kJ	kJ	kJ	kJ	kJ	kJ	kJ	kJ
Heating	1	88	891,881	1,821,588	−2,870,505	0	125,976	28,068	587	1,577	829
	2	220	998,907	2,132,513	−3,727,571	0	430,896	144,768	2,567	13,158	4,762
	3	385	1,355,659	2,325,514	−4,698,436	0	689,417	275,326	6,346	35,274	10,900
	4	500	285,402	1,072,936	−1,870,181	0	333,947	141,368	9,761	20,597	6,170
	5	585	927,556	1,473,563	−5,212,981	0	1,787,576	840,803	12,747	131,773	38,964
	6	675	356,752	1,012,973	−1,641,511	0	111,701	118,214	16,315	19,766	5,789
	7	765	927,556	1,620,031	−2,625,791	0	0	0	20,370	44,805	13,029
	8	895	927,556	1,652,681	−2,691,219	0	0	0	26,919	65,235	18,828
	9	998	535,128	1,145,442	−1,778,445	0	0	0	32,791	50,556	14,527
	10	1,083	677,829	1,140,434	−1,979,005	0	0	0	37,921	95,474	27,347
	11	1,155	356,752	688,205	−1,216,957	0	0	0	79,081	72,271	20,648
	12	1,190	142,701	290,794	−596,134	0	0	0	81,624	63,030	17,985
	13	1,203	35,675	85,837	−278,049	0	0	0	82,459	57,638	16,439
	14	1,206	7,135	−175,712	−13,220	0	0	0	82,654	77,143	22,000
Cooling	15	1,153	−756,315	−1,173,917	0	1,501,170	168,695	142,825	78,927	30,033	8,581
	16	1,055	−642,154	−1,151,100	0	1,123,403	403,493	155,280	70,582	31,473	9,024
	17	985	−356,752	−1,051,203	0	902,866	305,875	107,903	63,948	21,253	6,110
	18	855	−1,498,360	−2,369,133	0	2,188,839	1,132,546	399,528	51,047	74,102	21,432
	19	690	−856,206	−1,853,936	0	1,823,721	457,819	323,009	35,098	54,541	15,952
	20	565	−927,556	−1,638,384	0	1,704,986	137,609	582,531	24,328	89,849	26,637

(continued)

Table 5.27 (continued)

Stage	$T_{average}$	Q_{ware}	Q_{load}	$Q_{combustion}$	$Q_{cooling\ air}$	$Q_{difussion\ air}$	$Q_{injection\ air}$	$Q_{side\ walls}$	Q_{roof}	Q_{floor}
	°C	kJ	kJ	kJ	kJ	kJ	kJ	kJ	kJ	kJ
21	475	−356,752	−1,070,860	0	1,111,184	176,707	106,863	13,090	15,195	4,574
22	380	−998,907	−1,761,015	0	2,189,390	414,215	125,248	10,194	15,938	4,935
23	258	−749,180	−1,577,183	0	1,589,708	562,236	148,755	5,460	15,075	5,129
24	163	−606,479	−1,652,291	0	1,713,131	420,426	111,235	2,409	8,164	3,405

5.2.5.2 Global Energy Balance

Table 5.27 provides the values referred to the individual balance for each cycle stage. When combined, it can be obtained the global balance for the whole kiln as shown in Table 5.28.

The total sum of inlet and outlet enthalpies includes chemical and physical values of enthalpy referred to normal conditions. On the contrary, the Q values of energy represent how the heat value of the fuel is distributed for each thermal requirement within the kiln.

5.2.5.3 Residual Heat Throughout the Stack

The total amount of flue gases through the exhausting stack is shown in Table 5.29.

Potential heat recovery can be summarized as stated in Table 5.30.

49.5 GJ from residual heat means 92% of the total inlet enthalpy. Nevertheless, the temperature varies along the complete cycle. This fact hinders a feasible heat recovery as shown in Fig. 5.6.

The plot (a) shows a steadier curve of energy with a sharped peak about stage 5. However, when the energy is translated into power averaged by stage time, the peak is transferred to stages 15–17. Anyway, there are systems that may storage this residual power in order to stabilize a regular consumption.

5.3 Water Consumption Model

As explained before, water is required for different stages within sanitary-ware production. Water is essentially used as the raw material for slip, glaze and plaster molds production but it is also needed for flushing the molds after every casting. Depending on the water supply quality, it is common to treat the water removing dissolved salts, which is an important inconvenience for

- proper molds operation as a filtering media
- shaping of the slip in order to build up the piece inside the mold.

Resin molds technology is quite more intensive than plaster type in terms of water usage. Since the process does not involve the contamination of the water besides inorganic solid suspension particles and dissolved salts, it can be easily treated to be rejected to the sewage. On the other hand, the water containing salts is not easily treatable to reuse since its salt concentration increases with every cycle. Thus, water would degenerate progressively for production purposes.

The most common way to remove salts from water is reverse osmosis when it is intended to be reused. Otherwise, it is just wasted.

The equation system to assess the water consumption is by the analysis of processes, which include water in their operation and the subsequent mass balance.

Table 5.28 Balance for a shuttle kiln

In	H_{in}	m_{in}	Out	H_{out}	m_{out}	DH	Q_{fuel}	$Q_{required}$	%in	%out
	kJ	Kg		kJ	kg	kJ	kJ_{in}	kJ_{out}		
Ware	178,376	6,314	Ware	856,206	6,314	Q_{ware}		677,829		1.32
Plates	62,911	1,724	Plates	301,973	1,724	Q_{load}		553,906		1.08
Columns	10,948	300	Columns	52,548	300					
Supports	71,906	1,970	Supports	345,149	1,970					
Refractory concrete cart	54,466	2,476	Refractory concrete cart	168,901	2,476	Q_{cart}		433,870		0.85
Refractory panel cart	7,685	366	Refractory panel cart	23,709	366					
Refractory fiber cart	9,472	344	Refractory fiber cart	28,947	344					
Vermiculite 1/8 cart	101,720	16,953	Vermiculite 1/8 cart	291,473	16,953					
Vermiculite 1/5 cart	18,406	3,068	Vermiculite 1/5 cart	49,008	3,068					
Steel plate cart	14,576	1,267	Steel plate cart	38,376	1,267					
Wheels cart	91,080	7,920	Wheels cart	130,863	7,920					
CH_4 Heat value	51,310,449	974				Q_{NG}	– 51,310,449		100.00	
CH_4 Physical enthalpy	53,491		H_{CO2} OUT	1,304,554		Q_{exh}		15,761,647		30.72
O_2 combustion air	89,213	3,894	CO_2	7,560,389	2,677					
N_2 combustion air	333,110	12,818	H_2O from combustion	835,754	2,190					
O_2 excess air	8,921	389	Excess air	0	1,671					
N_2 excess air	33,311	1,282	N_2	6,578,997	12,818					
Cooling air	882,685	38,529	Cooling air	19,267,724	38,529	Q_{ca}		18,385,039		35.83
Diffusion air	513,965	19,778	Diffusion air	9,890,399	19,778	Q_{da}		9,376,435		18.27
Injection air	203,311	8,098	Injection air	4,049,894	8,098	Q_{ia}		3,846,583		7.50

(continued)

Table 5.28 (continued)

In	H_{in}	m_{in}	Out	H_{out}	m_{out}	DH	Q_{fuel}	$Q_{required}$	%in	%out
	kJ	Kg		kJ	kg	kJ	kJ_{in}	kJ_{out}		
			H_{walls}	847,224		Q_{walls}		847,224		1.6
			H_{roof}	1,103,921		Q_{roof}		1,103,921		2.15
			H_{floor}	323,994		Q_{floor}		323,994		0.63
Total In	54,050,003	128,465	Total Out	54,050,003	128,465	Balance	– 51,310,449	51,310,449	100.00	100.00

Table 5.29 Flue gases and air from a shuttle kiln

Exhausting				
Stack	Fluid	*m*	*T*	*Q*
		kg	K	kJ
Exhausting	CO_2	2,677	755	1,304,554
	H_2O	2,190	755	7,560,389
	Excess air	1,671	755	835,754
	N_2	12,818	755	6,578,997
	Cooling air	38,529	755	19,267,724
	Diffusion air	19,778	755	9,890,399
	Injection air	8,098	755	4,049,894
Residual heat		85,763	755	49,487,712

Table 5.30 Recoverable heat

Exhausting per species					
Fluid	*M*	*T*	*c*	*h*	*Q*
	kg/s	K	p.u.	kJ/kg	kJ
CO_2	2,677	755	0.0312	487	1,304,554
H_2O	2,190	755	0.0255	3,452	7,560,389
N_2	12,818	755	0.1495	513	6,578,997
Excess air	68,077	755	0.7938	500	34,043,772
Recoverable heat	85,763	755	1.0000	577	49,487,712

Fig. 5.6 Residual heat of a shuttle kiln

- *Water for slip production is calculated based on the composition formula of the ceramic slip which is about one third of the total wet weigh of the piece as described in Sect. 4.2 by Table 4.8.*
- *Water for glaze production is estimated from the assumption of the glaze/slip ratio is 1–9 kg of the final piece as described in Sect. 4.3 by Table 4.11.*
- *Water for plaster molds making is calculated from the number of molds produced daily as stated in Sect. 4.4.1.2 by (4.20).*

Table 5.31 Water requirement

Water consumption	Plaster molds factory				Resin molds factory			
	Process	Washing	Total day	%	Process	Washing	Total day	%
	m^3/day	m^3/day	m^3/day		m^3/day	m^3/day	m^3/day	
Raw materials		0.20	0.2	0.14%		0.20	0.2	0.02%
Slip preparation	25.57	0.50	26.07	18.37%	25.57	0.50	26.07	2.49%
Glaze preparation	2.84	0.50	3.34	2.35%	2.84	0.50	3.34	0.32%
Modeling and molds	8.93	0.30	9.23	6.50%		0.30	0.30	0.03%
Casting	26.94	2.00	28.94	20.39%	483.05	2.00	485.05	46.41%
Heating and ventilation								
Drying								
Glazing		2.00	2.00	1.41%		2.00	2.00	0.19%
Kilns								
Sorting		0.20	0.20	0.14%		0.20	0.20	0.02%
Offices	0.38	0.10	0.48	0.33%	0.38	0.10	0.48	0.05%
Ware house		0.25	0.25	0.18%		0.25	0.25	0.02%
Facilities	69.23	2.00	71.23	50.18%	525.35	2.00	527.35	50.45%
Total day	133.89	8.05	141.94	100%	1,037.19	8.05	1,045.24	100%
Annual consumption	31,062	1,868	32,930	m^3/year	240,628	1,868	242,496	m^3/year

- *Water for molds flushing within casting is quite different depending on the casting technology as described in Sect. 4.5.2.1 by (4.25).*
- *Water for sanitary usage is estimated per the common standard ratios for basins, showers,* etc.
- *With the exception of sanitary water, the rest has to be treated to remove salts. 50% efficiency is assumed for the R.O [2].*

The total water requirements of water are shown in Table 5.31.
In Fig. 5.7, the water assessment can be seen for a typical sanitary-ware factory.

5.4 CO_2 Emissions Model

To assess CO_2 emissions, both, electricity and thermal energies are taken into consideration to calculate their associated emissions [3] considering different indexes or affecting drivers [4]. For natural gas burnt, the stoichiometric rate of CO_2 is directly

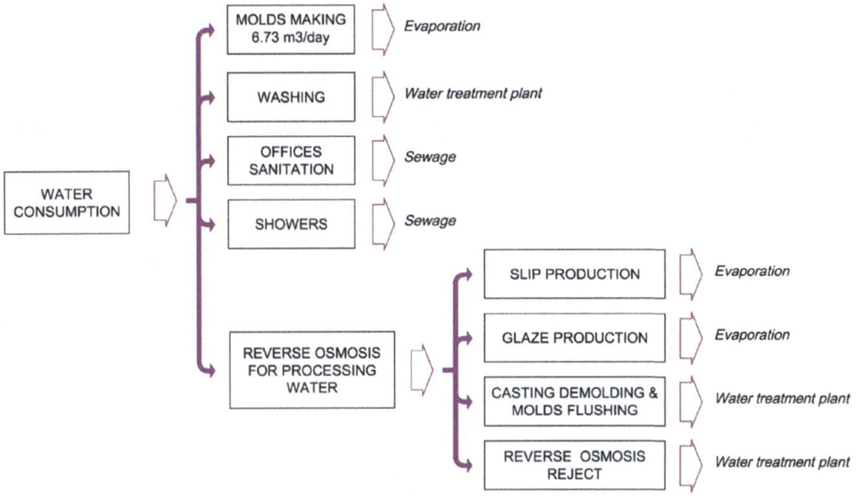

Fig. 5.7 Water consumption scheme

calculated from the natural gas that has to be burnt to produce the needed thermal power as follows:

$$1\ \text{kJ}_{\text{thermal}} \cdot \frac{1\ \text{kg}\ CH_4}{48,000\text{kJ}} \cdot \frac{1\ \text{kmol}\ CH_4}{16\ \text{kg}\ CH_4} \cdot \frac{1\ \text{kmol}\ CO_2}{1\ \text{kmol}\ CH_4} \cdot \frac{44\ \text{kg}\ CO_2}{1\ \text{kmol}\ CO_2}. \tag{5.4}$$

For the CO_2 produced by electrical power, it is calculated at a 35% efficiency of electrical production from a fossil fuel-based power plant upstream of the factory. Thus, the factor to evaluate the CO_2 production is:

$$1\ \text{kJ}_{\text{electrical}} \cdot \frac{1}{0.35} \cdot \frac{1\ \text{kg}\ CH_4}{48,000\text{kJ}} \cdot \frac{1\ \text{kmol}\ CH_4}{16\ \text{kg}\ CH_4} \cdot \frac{1\ \text{kmol}\ CO_2}{1\ \text{kmol}\ CH_4} \cdot \frac{44\ \text{kg}\ CO_2}{1\ \text{kmol}\ CO_2}. \tag{5.5}$$

5.5 Global Results

Once the models have been established and computed, the global results of the factory consumption can be summarized as per Table 5.32.

Figure 5.8 shows the final energy and water balance for the two types of factories studied.

After a detailed analysis of the factory requirements, it can be deduced that great amounts of energy (mainly thermal) and water are demanded to produce sanitary ware. Besides, the exhaust residual heats carry enough energy to be reused within the

Table 5.32 Compared consumptions and emissions for both types of factory

Consumption	Plaster mold factory				Resin mold factory			
	Electricity	Gas	Water	CO_2	Electricity	Gas	Water	CO_2
	kWh/year	kWh/year	m^3/year	t/year	kWh/year	kWh/year	m^3/year	t/year
Raw materials	435,928		46	257	435,928		46	257
Slip preparation	1,060,333		6,049	625	1,060,333		6,049	625
Glaze preparation	362,413		775	214	362,413		775	214
Modeling and molds	580,145	2,004,792	2,141	755	56,196		70	33
Casting	1,701,256		6,714	1,003	915,457	1,118,246	112,533	770
Heating and ventilation	183,524	4,351,851		1,006	183,524	609,956		234
Drying	565,732	2,112,524		769	565,732	2,112,525		769
Glazing	238,554		464	141	238,554		464	141
Kilns	935,650	18,800,768		4,429	935,650	18,800,768		4,429
Sorting	97,300	240,000	46	107	97,300	240,000	46	107
Offices	119,538		110	70	119,538		110	70
Ware house	35,554		58	21	35,554		58	21
Facilities	1,353,331	600,416	16,525	921	1,271,319	600,416	122,344	873
Total=	7,669,258	28,110,351	32,930	10,317	6,277,498	23,481,912	242,496	8,542

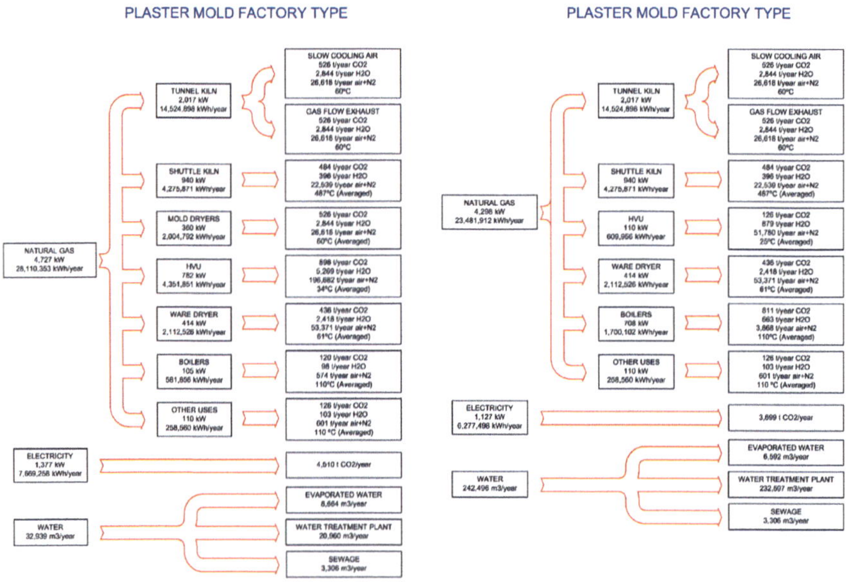

Fig. 5.8 Global energy and water assessment of the factory consumptions

Table 5.33 Examples of consumption in different plants

	Electricity	Gas	Water
	MJ/kgproduct	MJ/kgproduct	kgwater/kgproduct
Analytical estimation=	2.62	12.61	3.08
Plant 1 (Table 1.2) =	0.36	30	–
Plant 2 (Table 1.2) =	3.32	22	–
Plant 3 (Table 1.2) =	3.16	28	–
Plant 4 (Table 1.3) =	1.56	13.06	15.6

production processes, yet their enthalpy level is adjusted to be sequentially recovered in a thermal cascade [5].

It has been shown in Figs. 5.7 and 5.8, the amount of energy and water needed for both types of factories as well as the emissions produced by them. These quantities define one single factory sized at half a million pieces per year. The results show that pressure mold technology is more efficient from the energetic point of view and from the perspective of the emissions. For water consumption, pressure mold technology is not clean enough unless the water to flush molds can be reused.

The map of energy consumptions of the two main types of sanitary-ware factories is shown in Fig. 5.8 [6].

These analytical calculated results may be compared with actual values of the existing facilities as shown in Table 5.33.

References

1. Colmenar-Santos A, Terán de Lober LN, Borge-Diez D, Castro-Gil M (2013) Solutions to reduce energy consumption in the management of large buildings. Energy Build 56:66–77. https://doi.org/10.1016/j.enbuild.2012.10.004
2. Fritzmann C, Löwenberg J, Wintgens T, Melin T (2007) State-of-the-art of reverse osmosis desalination. Desalination 216(1):1–76. https://doi.org/10.1016/j.desal.2006.12.009
3. Zhou DQ, Wang Q, Su B, Zhou P, Yao LX (2016) Industrial energy conservation and emission reduction performance in China: a city-level nonparametric analysis. Appl Energy 166:201–209. https://doi.org/10.1016/j.apenergy.2015.09.081
4. Shao S, Yang L, Gan C, Cao J, Geng Y, Guan D (2016) Using an extended LMDI model to explore techno-economic drivers of energy-related industrial CO2 emission changes: a case study for Shanghai (China). Renew Sustain Energy Rev 55:516–536. https://doi.org/10.1016/j.rser.2015.10.081
5. Peng ZW, Liu ZJ (2009) Energy-saving and emission reduction for ceramic sanitary ware based on computer aided engineering. In: 2009 3rd international conference on bioinformatics and biomedical engineering, June 2009, pp 1–4. https://doi.org/10.1109/icbbe.2009.5162774
6. Cuviella-Suárez C, Colmenar-Santos A, Borge-Diez D, López-Rey Á (2018) Management tool to optimize energy and water consumption in the sanitary-ware industry. J Clean Prod 197, Part 1:280–296. https://doi.org/10.1016/j.jclepro.2018.06.195

Chapter 6
Improvement Proposals

In Chap. 4, the way to model the different consumers and the calculation of their consumptions were described. In Chap. 5, the results of these calculations were analyzed. From the available derived data set, it can be configured the energy-saving strategy by analyzing every consumption and the possibilities to cut it down.

In this chapter, they have explained the improved proposals from a qualitative point of view. In Chap. 7, they will be applied and computed in order to present results and savings.

There are many opportunities for saving energy within the processes along the production line. Mainly, they can be mentioned:

- Improvement of electrical efficiency involves many different possibilities. These actions could easily save up to 10% of the electrical consumption for lighting [1], robots [2] or compressors [3].
- Re-direction of kiln heat exhaust to be recovered at dryers and heating-ventilation units. This action could amount up to one third of the total energy consumption [4]. From Fig. 4.22, it can be deduced that, roughly, the exhausted heat from the kiln (1,681 kW) would be enough to supply the rest of the heat demands within a plaster molds factory. The total demand besides the kilns themselves reaches 1,918 kW (dryers and HVU's) in the plaster factory. The kiln is not able to supply completely this demand but the 88%. In a resin factory, the thermal for thermal consumers equals 729 kW that may be supplied from kiln exhausting totally. Some research has been done about this matter [5]. Even, the heat from the kiln exhaust could be redirected to preheat combustion air. Through this configuration, the kiln gas consumption would be greatly reduced. In the event that there was excess heat during some periods, it could be stored for later use [6]. Additionally, CO_2 emissions get reduced proportionally to energy decrease. This would be another source of economical savings [7] supported with real data [8], since CO_2 emissions may be punishable by local regulations.
- As there is recovered heat available, the treatment of the water could be dealt by distillation rather than reverse osmosis. This action would save electricity and water since salts would be removed and more than a half of the water would

C. Cuviella-Suárez et al., *Water and Energy Use in Sanitary-ware Manufacturing*, Green Energy and Technology, https://doi.org/10.1007/978-3-030-72491-7_6

be reused [9]. On the other hand, evaporated water can be condensed with an absorption chiller.
- Solar power is a truly effective efficiency tool that has been fostered by governments and, in fact, there are many solar plants supplying facilities at current, either for electrical [10] or thermal savings [11].
- Total power plant of cogeneration is rather advisable under these circumstances of energy consumption. In the event of a combined plant design, heat from the generator could be shared to be stored [12] or recycled to thermal processes. Total power plant allows for a better electrical load regulation [13] and re-use of residual heat [14]. Considering a total power plant of co-generation, good part of electrical supply might be generated on site with the benefit of an exhausting temperature over 450 °C from the generator. This amount of thermal energy could be completely reused throughout factory [15].

All these possibilities for saving make an interesting subject to be studied in detail to define an ideal factory for sanitary ware where energy and water consumption are minimized.

6.1 Generalized Proposals

In this section, generalized electrical reductions are analyzed to try to decrease the global electrical consumption. In Sect. 5.1, electrical consumptions were computed for every single consumer. Improvement actions have been analyzed individually in every section of the factory depending on the nature of the consumer. Nevertheless, some of the improvements are globally applicable to any kind of consumer or section all around the factory. Such is the nature of actions regarding:

- Electrical efficiency of the motors
- Speed control by frequency converters
- Substitution of conventional by LED lighting
- Lighting regulation by control of presence and external lighting level.

These actions are explained in the following paragraphs.

6.1.1 Efficiency of the Motors

The difference of electrical efficiency for different standardized levels (IEC-60034) can be up to 3% for common powers as can be seen in the following picture (Fig. 6.1).

Where:

- $IE1$ = standard efficiency
- $IE2$ = high efficiency
- $IE3$ = premium efficiency

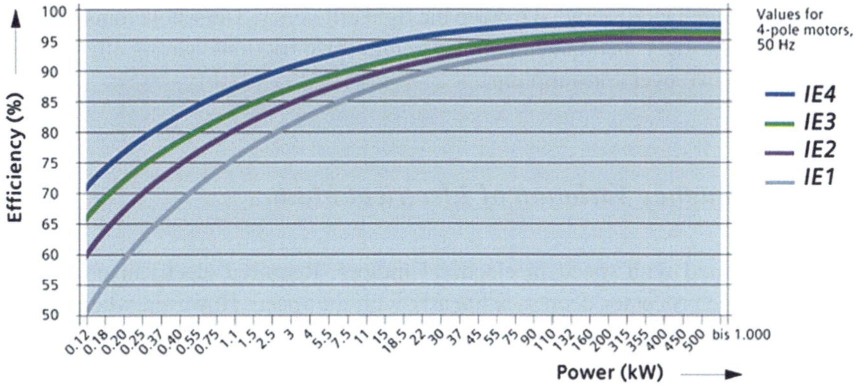

Fig. 6.1 Standardized levels of electrical motors efficiency (*Source* Data compiled by the authors)

- *IE*4 = super premium efficiency.

According to this difference and the annual influence, they can be computed energy saving and corresponding pay-back.

These IE-codes allow governments to specify the efficiency levels for their minimum energy performance standards (MEPS), which is set as *IE*2.

Let us assume that it is decided to improve efficiency of a standard *IE*2 motor by changing it into a *IE*3 one. The inlet power for the last has to be related with the former as:

$$P = P_{IE2} \cdot \eta_{IE2} = P_{IE3} \cdot \eta_{IE3}. \tag{6.1}$$

With the result of a power saving that equals:

$$\Delta P = P_{IE2} - P_{IE3} = P_{IE2} \cdot \left(1 - \frac{\eta_{IE2}}{\eta_{IE3}}\right) \tag{6.2}$$

and the annual saving

$$A = \Delta P \cdot N_h \tag{6.3}$$

where:

- $P_{IE2}: Electrical\ power\ motor IE2$
- $P_{IE3}: Electrical\ power\ motor IE3$
- $\eta_{IE2}: Electrical\ efficiency\ motor IE2$
- $\eta_{IE3}: Electrical\ efficiency\ motor IE3$
- $\Delta P: Power\ decrease$
- *A: Annual saving of energy*
- $N_h: Annual\ number\ of\ hours\ operation.$

Proper maintenance is crucial to keep the right efficiency. The whole transmission system has to be duly maintained in order to minimize frictions' wrong alignments, etc., which derive over consumption.

6.1.2 Frequency Variation of Electrical Motors

Torque is related with speed in electrical motors. Required electrical power for constant torque processes decreases linearly with the speed. However, when torque is depending on the square of the speed, the power decreases with the third power of the speed. Hence, electrical power can be reduced by decreasing speed as far as the process allows to do it without compromising the quality (Fig. 6.2).

Resistant constant torque:

$$P_1 = M_r \cdot n_1 = cte \cdot n_1 \tag{6.4}$$

$$P_2 = M_r \cdot n_2 = cte \cdot n_2 \tag{6.5}$$

$$\Delta P = cte \cdot n_1 - cte \cdot n_2 = cte \cdot n_1 \cdot \left[1 - (1 - \Delta n)^1\right] = P_1 \cdot \left[1 - (1 - \Delta n)^1\right]. \tag{6.6}$$

Resistant torque proportional to square of speed:

$$P_1 = M_r \cdot n_1 = cte \cdot n_1^3 \tag{6.7}$$

$$P_2 = M_r \cdot n_2 = cte \cdot n_2^3 \tag{6.8}$$

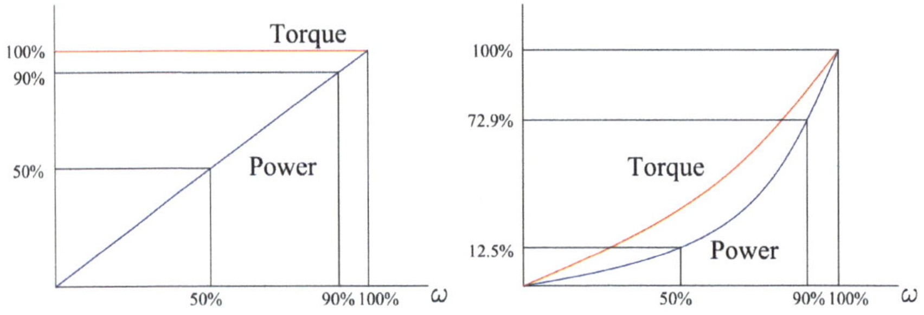

Fig. 6.2 Torque and power variation with the speed. **a** Constant torque. **b** Torque proportional to second power of speed (*Source* Data compiled by the authors)

$$\Delta P = cte \cdot n_1^3 - cte \cdot n_2^3 = cte \cdot n_1 \cdot \left[1 - (1 - \Delta n)^3\right] = P_1 \cdot \left[1 - (1 - \Delta n)^3\right]. \tag{6.9}$$

Thus

$$A = \Delta P \cdot W_{\text{year}} \tag{6.10}$$

where:

- A : *Ahorro anual*
- M_r : *Momento resistente*
- P_1, P_2 : *Potencias con y sin variador*
- n_1, n_2 : *Velocidades con y sin variador*
- Δn : *Disminución de velocidad porcentual*
- $W_{a\ no}$: *Energía anual a la velocidad nominal.*

6.1.3 Substitution of Conventional by LED Lighting

LED lighting is quite more efficient than conventional, therefore, a great chance of saving is related to this fact. The range of specific consumption is relatively wide depending on the technology and quality of the light fitting. To compute the potential saving, they have been considered following parameters (Table 6.1).

Thus, assuming same flux with both technologies:

$$Lighting = P_{\text{con}} \cdot \eta_{\text{con}} = P_{\text{led}} \cdot \eta_{\text{led}} \tag{6.11}$$

$$\Delta P = P_{\text{con}} - P_{\text{led}} = P_{\text{con}} \cdot \left(1 - \frac{\eta_{\text{con}}}{\eta_{\text{led}}}\right) \tag{6.12}$$

$$A = \Delta P \cdot N_h \tag{6.13}$$

where:

- r_s : *Retorno simple*
- P_{n_1} : *Potencia iluminación con alumbrado convencional*

Table 6.1 Comparative consumption of lighting lamps. (*Source* Data compiled by the authors)

Lighting efficiency	*h*	
	lm/W	
	Fluorescence	Discharge
Conventional	80	120
LED	100	140

- P_{n_2} : *Potencia iluminación con alumbrado LED*
- N_h : *Número horas.*

6.1.4 Lighting Regulation by Control of Presence and External Lighting Level

Lighting requirement is directly related to both, human presence and available external or natural lighting. In this regard, passive architectural design plays a very important role when factory is designed in order to allow for day light as long as possible depending on the geographical location. On the other hand, lighting has to be restricted to human presence under automatic operation to be run only when needed. Referring to day light, a proper lighting regulation system can save much energy along the year adjusting the light level to be added to the available day light level at any moment of the day. Thus, sunny days are less demanding on electrical power than cloudy ones. In the same way, dawn or dusk hours require some electrical light level less than 100% since a fraction of day level is available. Two coefficients have to be defined to estimate the fraction of time there is the human presence and there is available some fraction of day light.

Hence, the sized and effective electrical power for whole year are computed as follows:

$$C_{\text{reg}} = P \cdot N_h \cdot p_{el} \cdot \left(1 - f_{\text{presence}}\right) \cdot \left(1 - f_{\text{exterior}}\right) \tag{6.14}$$

$$A = C - C_{\text{reg}} = C \cdot \left[1 - \left(1 - f_{\text{presence}}\right) \cdot \left(1 - f_{\text{exterior}}\right)\right] \tag{6.15}$$

where:

- C, C_{reg} : *Coste anual sin y con regulación*
- $P : Potencia$
- N_h : *Número de horas anuales*
- $p_{el} : Precioelectricidad$
- C_{reg} :*Coste anual con regulación*
- $f_{\text{presence}} : factordeahorroporpresencia$
- f_{exterior} :*factor de ahorro por iluminación diurna.*

6.2 Particularized Actions Per Section

Once generalized actions have been described to be applied to any electrical device, a group of particularized measures are described in the following sections regarding different consumers whose special characteristics of operation allow specific saving actions. On the other hand, some general actions are described, which may be implemented to increase the efficiency of the global process as for instance; alternative energies, combined heat and power plants.

The following sections through the manufacturing line:

- Glazing
- Sorting
- Offices
- Ware house.

Do not present any particular characteristic to allow specific actions further than generalized ones. Thus, they are not mentioned in this section.

Glazing preparation section is neither mentioned since its configuration equals slip preparation with the only difference of dimension. All said for slip applies to glaze in a lower scale factor.

6.3 Raw Materials

6.3.1 Mills

In Sect. 4.1, it was stated the torque that mil has to exert to roll the load as a function weigh and rotational angle as:

$$T = r_g \cdot mg \cdot sen\alpha + r \cdot \mu \cdot mg \cos\alpha. \tag{6.16}$$

When mill is rolling, the load tends to raise internal load until centrifugal force is equaled by the radial component of the weigh that pebbles start to fall down as Fig. 6.3 shows.

The higher the speed, the higher rotational angle for the load. The equation for this equilibrium yields the function $\alpha = f(\omega)$ as follows:

$$m \cdot r \cdot \omega^2 = mg \cdot sen\alpha. \tag{6.17}$$

Combining both equations it can be derived a function of torque depending on speed $T = f\ (\omega)$ as:

$$T = r_g \cdot mg \cdot \frac{r\omega^2}{g} + r \cdot \mu \cdot mg \cdot \sqrt{1 - \left(\frac{r\omega^2}{g}\right)^2}. \tag{6.18}$$

The behavior of this function can be assumed as a second-order polynomial. Figure 6.4 shows both curves that are overlapped. Consequently, electrical power is very approximately decreased with the second power of speed decrease.

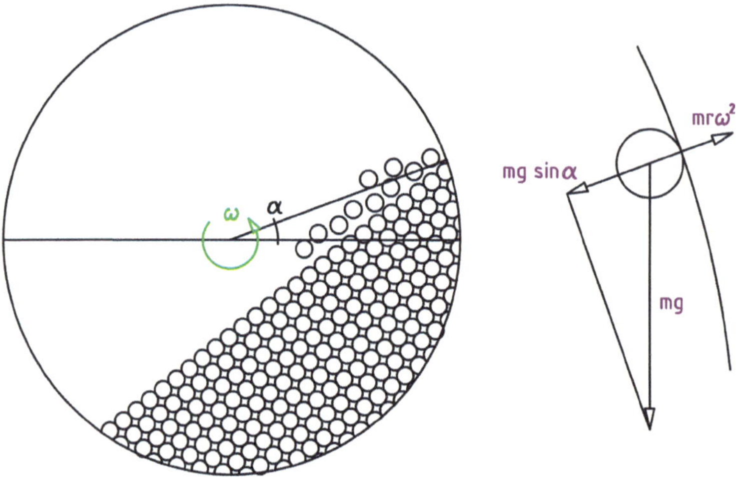

Fig. 6.3 Cascade effect of the mill operation

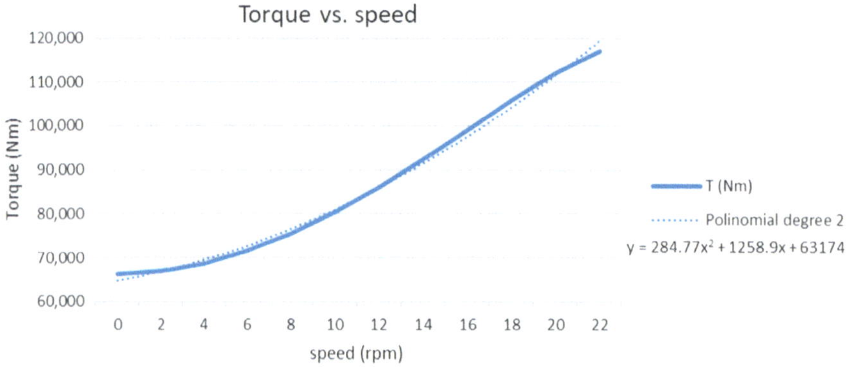

Fig. 6.4 Graph of function torque = f(speed) for the mill

There is another easy way to minimize the required power to run the mill. Optimal operation of the mill requires 100% of designed load in order to minimize torque. As per Fig. 6.5, torque depends on the radius of the center of mass, r_g.

According to:

$$M_r = r_g \cdot mg \cdot sen\alpha. \tag{6.19}$$

When mill is not full, the resistant torque increases and so does the required power. Thus, it an advisable practice to prepare milling batches matching the full capacity of the mill. It does not require investment and production cost would not increase significantly.

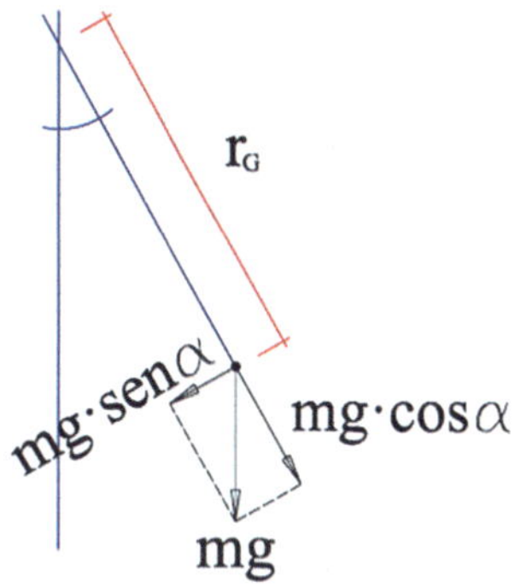

Fig. 6.5 Center of masses for displaced load

6.3.2 Belts, Sieves, Magnets and Other Equipment

As electrical equipment, general reduction actions are applicable to all of them.

6.4 Slip Preparation

6.4.1 Blungers

Power is about 125 kW, which represents a good part of the consumption in this section. Thus, it is advisable to keep right efficiency and upgrade EE degree when possible. The power for this equipment is devoted to move liquid slip so resistant torque is proportional to the square of the speed. Hence, it is very advisable to apply a reduction of speed as far as the process allows without a defect of quality. The reduction of power is proportional to the third power of the speed increase.

6.4.2 Tanks for Stock of Slip

All the measures recommended for blungers are valid for this equipment since the electrical behavior is quite like the same with only difference in power.

6.4.3 Peristaltic Pumps, Belts, Sieves and Other Equipment

As electrical equipment, general reduction actions are applicable to all of them.

Table 6.2 Operation parameters of a mold dryer

Air condition	T^a	HR	e	$X_{saturation}$	X
	°C	%	mbar	g/kg	g/kg
Ambient	22	55	26.45	16.674	9.17
Dryer chamber	60	75	199.37	152.42	114.31

6.5 Modeling and Molds

6.5.1 *Mold Dryers*

In this section, main consumption is located at mold dryers, so electrical as thermally. General electrical improvements are applicable, though frequency variation does not produce much impact since variability of speed is very tight.

Regarding thermal efficiency, it can be improved by means of different actions. Most thermal power in a mold dryer is devoted to evaporate the water contained within the plaster, which is not reducible. Notwithstanding, there are other consumptions that can be treated in order to reduce or eliminate them. Table 6.2 shows usual drying conditions during operation of a mold dryer.

Thermal consumption can be slit as follows:

- Air heating
- Water heating
- Water evaporation
- Plaster heating
- Carts/supports heating
- Conduction losses throughout enclosure.

The first four consumptions are physically defined by the required energy to evaporate the water content. Nevertheless, the efficiency of these processes is not perfect, so thermal requirements for them are improvable by some actions as well as the rest of the consumptions that are described below.

6.5.2 *Hygrothermal Regulation of the Dryer Environment*

Given the amount of water to evaporate and the air conditions in the chamber, it can be calculated the required amount of air to take this water out. For this purpose, a proper fan is designed to carry out the air extraction during the drying cycle. What usually happens is that fan capacity is a constant on the contrary of the air requirement, which is set by hygrothermal condition. As a result, an excess of air is blown into the chamber producing an extra heating requirement. By means of a proper regulation system, the air flow blown into the chamber would be exactly adjusted to that one

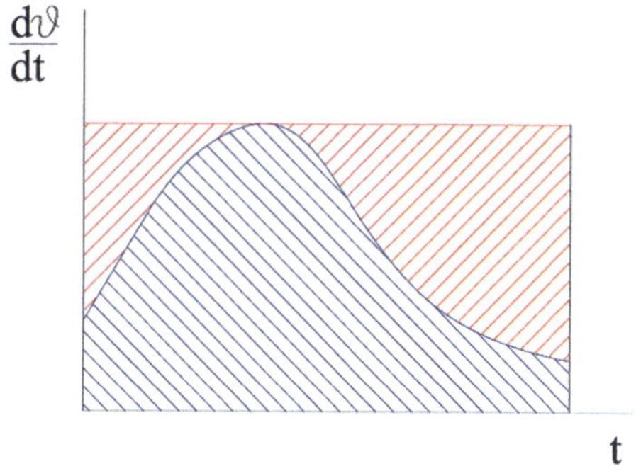

Fig. 6.6 Constant versus variable air flow

required by the dryer. Such system should consider a frequency converter for the fans and temperature and humidity sensors.

With a proper study of the specific drying curve of the plaster, it can be calculated the pace of air blowing during each stage of drying. As an example given, Fig. 6.6 shows a standard curve of drying rate.

The frequency variation of fans determines the exact air flow rate to match evaporating rate. Thus, a minimum of heat is required to heat up the inlet fresh air. Otherwise, air flow would be constant at the maximum required rate. Thus, the energy of the red areas can be saved. Electrical power to move the motors of the fans is as well reduced in the third power since the resistant torque is proportional to the square of the speed.

Another situation that can happen appears when running the dryer off load. The less the mass inside the dryer, the less required fresh air.

6.5.3 Insulation Improvement

Depending on the nature and condition of the existing insulation, the reduction of conduction and infiltration losses can reach 10% when the original insulation is not adequate. From the model of this book, the share of heat consumption within the mold dryer is as stated in Table 6.3.

No matter what is the global coefficient of thermal transmission of the dryer enclosure panel to conclude that it would be enough to double the thickness to cut the losses in a half (8 kW). This action provides a 5.5% reduction in thermal consumption.

Table 6.3 Distribution of thermal consumption in a mold dryer

Dryer Q_{demand}		
	kW	%
$Q_{supports}$	0.25	0.16
Q_{molds}	0.85	0.56
$Q_{vaporization}$	100.67	65.91
Q_{air}	13.68	8.96
$Q_{exhausting}$	20.66	13.52
Q_{losses}	16.64	10.89
	152.75	100

6.5.4 Recovered Heat

The clearest action is the thermal recovery of residual heats from kilns as it is already being done in some factories. This action would eliminate the total thermal consumption of the dryers depending on the rest of the thermal equipment that might be added to the recovered heat network.

6.6 Casting

6.6.1 Casting Benches

Electricity consumption in casting benches is important, although it is quite fractioned. The global value results from the sum of multiple small motors and lighting. It is worth studying the set of benches in the room to optimize unit and/or group consumption.

6.6.2 Transformer for Handling Device

Each line is usually equipped with a transformer to supply handling equipment. A larger transformer able to supply all the benches would reduce the losses of the alternative of individual transformer for each bench.

In this investment, the savings are almost negligible, but the price of the single transformer is less than that of 15 small transformers. From the energy point of view, a single transformer would be slightly more interesting than many of them distributed, but the disadvantages are that in the event of a breakdown, all production would stop. On the other hand, besides the saving on the price of the transformer must be added the supply network of the 30 benches from the single transformer.

Table 6.4 Regulation of fans to dry molds. (*Source* Data compiled by the authors)

Mold fans electrical regulation		
Number of fans	1,350	
Fan power	0.125	kW
Stop factor	0.5	
Drying cycle	16	H
Savings	313,200	kWh/year

6.6.3 Fans to Dry Molds

Each bench position is usually equipped with a centrifugal fan to dry the mold. This fan works continuously. Their joint consumption is high due to the number of units that exist, although each one represents less than 1 kW. Automatic management of its operation could reduce consumption. For this, a preliminary study must be carried out to determine the necessary operating times. An intermittent duty cycle or reduced operating hours can reduce consumption. The operation of the fans can be reduced by 50%. This is undoubtedly an interesting investment, but it requires empirically establishing the fan stop factor and the mode of operation of the fans without compromising the drying of the molds. As an example, assuming a stop factor of 0.5, saving would be (Table 6.4).

6.6.4 Local Lighting Lines

Further than general lighting distribution, every bench is equipped with by fluorescent lighting units. By substituting them with LED's units, electrical power might be reduced over 15%. Length of a bench is up to 45 m. Electrical power for it would reach up to:

$$\frac{58\ \mathrm{W}}{1.5\ \mathrm{m}} \cdot 45 \frac{\mathrm{m}}{\mathrm{bench}} \cdot 30\mathrm{benches} = 52.2\ \mathrm{kW}. \tag{6.20}$$

From Table 6.1 and (6.11)–(6.13), they can be calculated the savings for 1 year of operation as follows:

$$8\frac{\mathrm{h}}{\mathrm{day}} \cdot 232\frac{\mathrm{days}}{\mathrm{year}} \cdot 52.2\mathrm{kW} \cdot \left(1 - \frac{80}{100}\right) = 19,377\frac{\mathrm{kWh}}{\mathrm{year}}. \tag{6.21}$$

6.6.5 Pressure Machines

A pressure casting machine consists of multiple different electrical and pneumatic devices. Thus, all the general measures for electrical system are applicable to them.

6.6.6 Tanks for Stock of Slip

As explained for slip tanks in the slip preparation section, these tanks are provided with motors in operation clock around. Thus, it is advisable to improve efficiency of them so as to apply frequency converters in order to reduce the speed of agitation.

6.7 Heating and Ventilation of Casting Hall

6.7.1 Heating and Ventilation Units

Electrical power is led mainly by fans, either to blown air in or to extract it. Air handling is usually driven by large motors coupled to great fans, (up to 25 kW), on the other hand, there might be different smaller fans to extract air from the hall. Any mentioned action to reduce electrical consumption can be applied to them.

Regarding thermal consumption, it could get reduced by different ways:

6.7.2 Heat Recovery

The moderated degree of temperature required for heating allows to take profit of recovered heat from kilns exhausting. A properly treated mixture of heated air from kiln, fresh air and moisture can be blown into the casting hall instead of heating air by means of a gas operated unit. This action may reduce most of gas consumption in this section.

6.7.3 Hygrothermal Regulation

A suitable hygrothermal regulation can reduce the thermal needs by reducing the circulation of fresh air. The cycle load curve must be studied for the optimal drying of ware and molds, which is ultimately the objective of heating. The inlet, recirculation, heating and air extraction flows must be closely linked to each other and according to a load curve adapted to the specific needs of the room to be heated at any moment.

When this regulation does not exist and the size of the ventilation is not adequate, it can happen that there is an excess of ventilation and thermal consumption. A real energy balance must be carried out to calculate the potential reduction of consumption as explained in Sects. 4.6 and 5.22. In addition to this thermal power, it has to be taken into account the electrical saving derived from speed regulation.

6.7.4 Enclosure Insulation

Better insulation of the room would reduce conduction losses and could save between 5 and 10%. From Table 5.18, it has been stated the losses throughout enclosures for the different period. These values have been calculated with an overall coefficient of thermal transmission "U" of 0.52 $W/m^2/K$ and surface of 7,701 m^2. An improvement in insulation by making thicker the panel on the roof and walls would reduce the enclosure losses.

6.8 Ware Dryers

6.8.1 Thermal Parameters of Ware Dryer Operation

Drying parameters are shown in Table 6.5.

Distribution of thermal requirement for each phase of the drying process is shown in Table 6.6.

Table 6.5 Operation parameters of a ware dryer

		Ambient air	Heating 1	Heating 2	Heating 3	Heating 4	Cooling
States		0	1	2	3	4	5
Duration	h		2	3	2	3	1
Time	h	0	2	5	7	10	11
T	°C	25	38	50	75	90	25
RH	%	55%	75	50	30	20	55
Water to evaporate	%		2.50	3.5	5.00	6.00	
Water to evaporate	kg		539	754	1,078	1,293	
Fresh air for evaporation	kg		26,061	23,503	10,352	4,717	
Thermal requirement	kW		357	336	695	480	

Table 6.6 Distribution of thermal requirements in a ware dryer

Thermal requirement per phase						
	Heating 1	Heating 2	Heating 3	Heating 4	Cooling	Overall
$Q_{\text{fresh air}}$	45	55	72	29		52
$Q_{\text{fresh vapor}}$	1	1	1	1		1
Q_{NG}	0	0	0	0		0
$Q_{\text{combustion}}$	44	43	97	70		54
Q_{ware}	31	21	63	25		0
Q_{carts}	12	8	23	9		0
$Q_{\text{cooling air}}$	0	0	0	0	415	37
$Q_{\text{vaporization}}$	212	185	392	286		233
Q_{losses}	12	24	47	61	31	37
Total	357	336	695	480	446	414

Due to the similarity of operation, the same actions described for mold dryers in Sect. 6.5 are applicable to ware dryers even though the operation parameters are different when drying ware.

6.8.2 Insulation of the Enclosure

An increase of dryer insulation reduces conduction thermal losses. Depending on the nature and condition of the existing insulation, the reduction of conduction and infiltration losses can reach 10% if the original insulation is not adequate. The global coefficient of thermal transmission of the dryer's enclosure panel is 2.83 $kJ/m^2/K$. It would be enough to double the thickness to cut the losses in half (18 kW on average), which represents 4% of savings.

6.8.3 Regulation of Fresh Air Intake

As in mold dryers, a mismatch of the ventilation flow leads to excess thermal power. In this case, speed variation is even more justified since the ventilation flow is different in each drying phase. For the dryer, the ventilation rates and excesses are shown in the Table 6.7.

Actual intake of air when no frequency inverter is supplied for fan exceeds of needed and has to be heated up as expressed by Q_{excess}. This thermal power can be avoided by regulating the air intake flow at the required level.

Table 6.7 Thermal reductions by regulating ventilation. (*Source* Data compiled by the authors)

Savings by air intake regulation						
		1	2	3	4	Average
Air intake	kg/h	13,030	7,834	5,176	1,572	6,463
Actual intake	kg/h	13,030	13,030	13,030	13,030	13,030
Q_{excess}	kW	0	12	55	69	35

6.8.4 Heat Recovery

The required thermal power for the dryer might be supplied from heat recovery of the kiln exhausting. By means of this action, it could be saved the total amount fuel required by dryer as per Table 6.6 that means an average of 455 kW (cooling not considered for average).

6.9 Kilns

The kilns are responsible for up to 75% of thermal consumption of the factory, hence they require special attention, especially considering that approximately 66.70% of the heat generated in the kilns goes through the chimneys, as the following graph shows (Fig. 6.7).

The tunnel kiln has been modeled in this book to fire the load of 500.000 toilet seat per year with a design thermal power of 2,017 kW. Sanitary ware kilns have a standardized consumption of 5,225 kJ/kg:

The Table 6.8 shows the exhausting and air output data for it.

Table 6.9 shows the distribution of thermal requirements within the tunnel kiln.

Residual heat of the kiln may be reused either in the kiln itself or other thermal consumers as dryers, heating units, etc.

6.9.1 Induced Draught and Pre-Heating of Combustion Air

The residual heat from the stacks of the kiln can be used to preheat the combustion and roof airs in order to reduce the amount of gas proportionally to the increase of enthalpy of the mixed air. This can be done in two levels of temperature that can be independent, using the slow cooling air, with a high flow rate and relative low temperature and using the rapid cooling air with low flow and high temperature. Usually, rapid cooling air is already recovered inward due to the induced draught produced by the exhausting fan as shown in Fig. 6.8.

This flow is calculated by the needed amount of air to cool down the masses inside the tunnel before entering the rapid cooling section. It may happen that rapid cooling

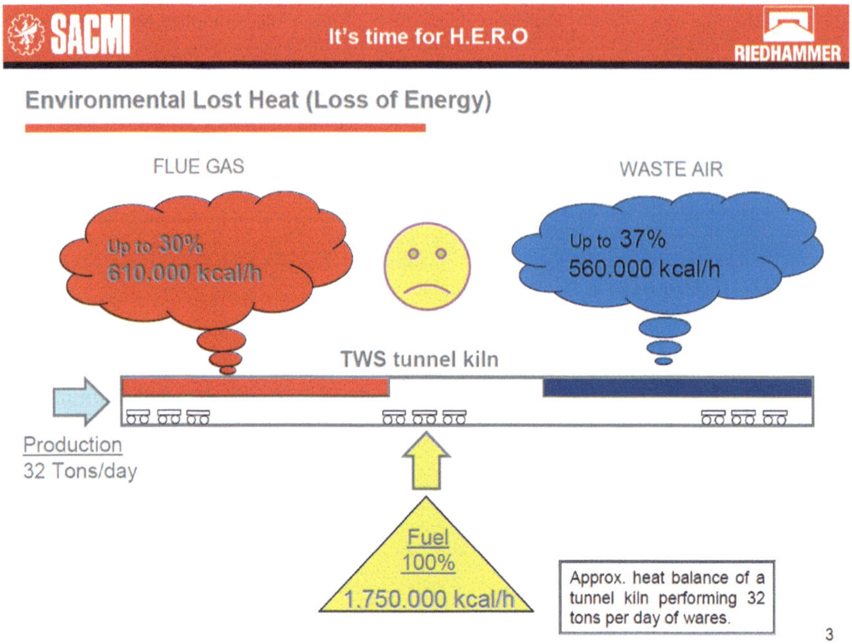

Fig. 6.7 Relationship between fuel and exhausting powers in an industrial SACMI kiln (Courtesy of SACMI GROUP)

Table 6.8 Distribution of thermal power through the stacks of the conventional kiln. (*Source* Data compiled by the authors)

Exhausting per Stack				
Stack	Fluid	*m*	*T*	*Q*
		kg/s	K	kW
Exhausting	HCO_2 OUT	0.1156	537	29
	HH_2O OUT	0.0946	537	284
	Hair excess OUT	0.0721	537	19
	HN_2 OUT	0.5534	537	153
	H_{roof} air OUT	0.3120	537	84
	$H_{draught}$ OUT	0.0749	537	20
	H_{rapid} cooling OUT	0.6553	537	176
Slow cooling	H_{slow} cooling OUT	2.9677	576	916
Recovered Heat		4.8455	561	1,681

Table 6.9 Distribution of the thermal requirements for the conventional kiln

Heating requirements		
Thermal load	kW	%
Q_{ware}	34	1.70
Q_{load}	31	1.55
Q_{cart}	84	4.18
$Q_{\text{exhausting}}$	477	23.67
$Q_{\text{roof air}}$	81	4.01
$Q_{\text{rapid cooling}}$	158	7.84
$Q_{\text{slow cooling}}$	842	41.74
Q_{losses}	309	15.32
Total	2,017	100.00

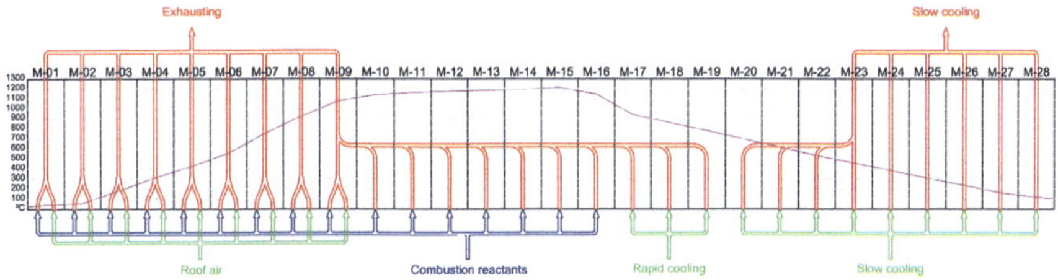

Fig. 6.8 Air flow circuits in a conventional tunnel kiln

air that has been already heated through rapid cooling section is not cool enough to drop down temperature in the last module of the firing section. In this case, the induced draught increases up to suction air through the chimney. It can be seen also, how the rapid cooling is providing fresh air inwards the tunnel in order to help the cooling action of the induced draught to cool down masses before entering rapid cooling section.

On the other hand, the slow cooling recovered air might be reused in different levels of temperature to pre-heat roof air as well as combustion air as per the ranged temperature in each module of firing section. As a result of this action, the final exhausting flow gets reduced as well as the slow cooling recoverable heat. Since cooling air is retrofitted to combustion section, the available recoverable heat from slow cooling residual air is lower than it was before preheating.

When combined both factors (induced draught and pre-heating), it results in an optimized kiln conceived as a heat counter-current heat exchanger as shown in Fig. 6.9.

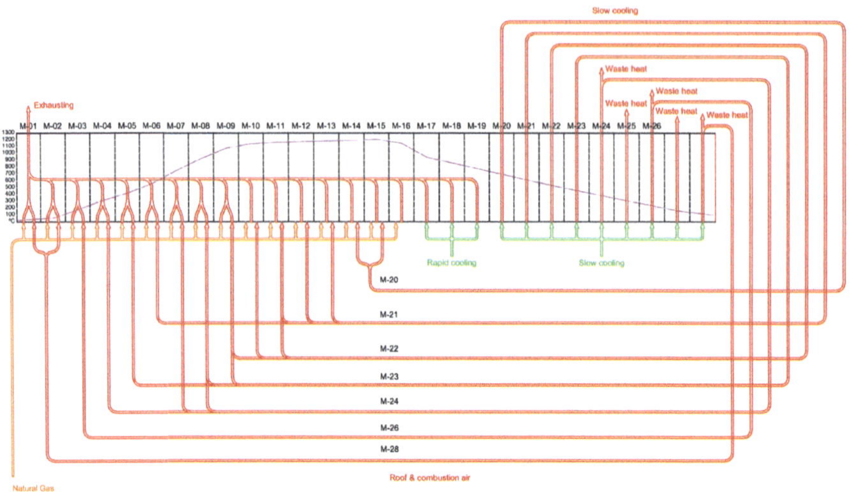

Fig. 6.9 Air flow circuits in an optimized tunnel kiln

6.9.2 *Pulse Burners*

This is a simple concept that allows for firing efficiency. Burners are designed to get the optimal efficiency by the full load capacity. When regulated to a lower load, efficiency is decreased. In this sense, the regulation is rather obtained by running and stopping burner at full load capacity during stated periods as per the desired averaged power. Besides, running burners are acted in the way to produce effective turbulences facilitating heat convection and increasing heat transmission.

Most of the current kiln manufacturers are developing systems whose purpose is a global control of combustion and heat transfer base on the following principles

- Pulsed burners in the pre-heating zone with pre-heated air at 200 °C
- Air pre-heated to 400 °C (EMS-400-4.0) or 700 °C (EMS-700-Cyber) in the heating zone.
- Controlled recovery system in the rapid cooling zone.
- Electronic air/gas ratio control
- Performance control combined with cooling airs

6.9.3 *Controlled Regulation of the Kiln*

Temperature curve of the kiln and fuel flow should be monitored with appropriate frequency in order to detect any failure within the system that may produce any extra consumption. This solution seems trivial and it really is, but the truth is that the kilns are not regulated with the precise frequency, so operation becomes not optimal. This causes air and fumes to move improperly within the kiln chamber, causing

Table 6.10 Common specific consumption of kilns. (*Source* Data compiled by the authors)

Kiln	Load	Specific consumption	
Load	kg/m^3	kcal/kg	kWh/kg
Tunnel kiln	150	1,260	1.46
Shuttle kiln	170	1,800	2.09

unnecessary heating of air flows. This happens particularly when the draft of the combustion fume extractor rises above what is necessary, dragging the cooling airs toward the combustion zone in excess. Under such circumstances, the firing zone may result in overcooled. Thus, extra fuel has to be supplied to keep the temperature inside the chamber.

6.9.4 Load Control

This is another trivial solution, which all factories are aware of. However, due to production necessities, it is not always practiced. An adequate kiln load is essential to optimize kiln performance, which may be briefed into two main principles:

- The load must consist mostly of product than refractory supports
- The supports must be of the lightest material and with the least heat capacity possible, that is, silicon carbide better than cordierite.

An organization of the production that allows to group products in a compact way on the cart would allow to reduce the relative quantity of refractory. When production lots are not well arranged, the ratio ware-refractory gets further than optimal. The Table 6.10 shows common load densities and typical warranty efficiencies as optimal ones for standard kilns.

6.9.5 Substitution of Cordierite by Silicon Carbide for Refractory Supports

The replacement of cordierite by SiC (Silicon carbide) generates a decrease in thermal consumption due to the conjunction of different factors. The density of SiC is greater than that of cordierite but its mechanical resistance is greater to a greater extent, as shown in the Table 6.11.

When changing from supports of cordierite into SiC ones, it can be approximated that the ratio of cross-sections is inversely proportional to MOR and so does the volumes ratio:

$$\frac{v_{\text{cordierite}}}{v_{SiC}} = \frac{MOR_{SiC}}{MOR_{\text{cordierite}}}. \tag{6.22}$$

Table 6.11 Comparative parameters of supporting materials. (*Source* Data compiled by the authors)

Refractory	Density	MOR	Ce
	kg/m^3	MPa	kJ/kg·K
Cordierite	2.6	200	1.46
SiC	3.2	540	0.92

Thus, the ratio of masses would be:

$$\frac{m_{\text{cordierite}}}{m_{SiC}} = \frac{v_{\text{cordierite}} \cdot \rho_{\text{cordierite}}}{v_{SiC} \cdot \rho_{SiC}}. \tag{6.23}$$

And finally, the ratio of thermal power required is:

$$\frac{Q_{\text{cordierite}}}{Q_{SiC}} = \frac{m_{\text{cordierite}} \cdot Ce_{\text{cordierite}}}{m_{SiC} \cdot Ce_{SiC}} = \frac{MOR_{SiC} \cdot \rho_{\text{cordierite}} \cdot Ce_{\text{cordierite}}}{MOR_{\text{cordierite}} \cdot \rho_{SiC} \cdot Ce_{SiC}} = 3.48. \tag{6.24}$$

6.9.6 *Insulation Increase*

Although conceptually better insulation reduces conduction losses through the walls, which do not represent 10% of consumption and given the difficulty of providing over-insulation in the tunnel closing walls due to the number of existing auxiliary elements, this measure is neither practical nor economical when dealing with an existing kiln. Nevertheless, it may be an advisable recommendation for a new kiln.

6.9.7 *Refractory Lining Film*

When heat transfer throughout enclosures is calculated, it is demonstrated how much losses through walls and roof are depending on the emissivity coefficient. Thus, a proper treatment to increase it makes the heat transfer getting lower.

6.10 Facilities

The electrical consumption of this section is mainly led by compressors and secondly by the pumping systems; water treatment, fires, etc.

The measures to be adopted in the case of compressors are widely known and extended. Below, they are described.

6.10.1 Frequency Variation

Depending on the compressor load, the frequency variation allows to reduce the cube of the power proportionally to the reduction of the operating regime. When there is a network made up of different compressors, what prevails is an automatic management system to operate them in stages and works with one of variable speed to adjust the load level. In the case of this study, the average power of compressed air is 211 kW and the size of it 300 kW. The demand may be supplied by either one or two compressors.

Let us assume a curve demand as the following figure shows for the cycle of 1 day (Fig. 6.10)

During the central hours of the shift, the maximum is presented and during the night hours, the system will work at minimums. The optimal distribution of power from the consumption point of view would be a constant speed compressor with 50% of the maximum power plus another variable speed compressor with 50%.

One compressor at constant frequency may supply up to 150 kW, operating in a constant basis of power. The rest of variable power may be supplied by a variable speed compressor, whose power is managed by the speed of the motor adjusting the consumption to the required at any moment. On the contrary, a constant speed compressor would take 150 kW all the cycle long.

The difference between both consumption results 2,141 kWh by integrating the curve for the whole cycle. The annual potential saving would be:

$$2,141\text{kWh} \cdot 232\frac{\text{days}}{\text{year}} = 496,712\frac{\text{kWh}}{\text{year} \cdot \text{compressor}}. \tag{6.25}$$

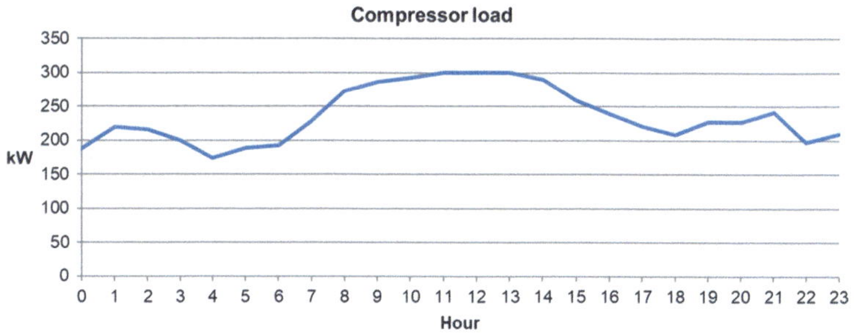

Fig. 6.10 Electrical demand of the compressor (*Source* Data compiled by the authors)

6.10.2 Heat Recovery from Compression Cooling

When air is compressed in a compressor, its temperature increases according to an adiabatic process according to:

$$P^{1-\gamma} \cdot (nRT)^{\gamma} = k. \tag{6.26}$$

When the final pressure is 8 bar, the adiabatic temperature of the air increases up to 276 °C. In order to adjust it to required conditions and to reduce the volume, there is provided a refrigeration system, which properly redirected, can return up to 80% of the power input as heat, depending on whether the cooling is by air or water.

6.10.3 Appropriate Maintenance of the Compressed Air Network

A proper network maintenance that eliminates any leak may produce significant savings. This factor is difficult to assess, but it can be made a rough estimation by Bernouilli's equation. Let us assume an 8 bar network with a small leakage of 5 mm diameter. The velocity of air leaking would be:

$$v = \sqrt{\frac{P \cdot 2}{\rho}}. \tag{6.27}$$

And the air flow:

$$q_{\text{air}} = v \cdot s = \sqrt{\frac{P \cdot 2}{\rho}} \cdot \pi \cdot \frac{D^2}{4} \sqrt{\frac{8\text{bar} \cdot \frac{10^5 Pa}{\text{bar}} \cdot 2}{1.225 \frac{\text{kg}}{\text{m}^3}}} \cdot \pi \cdot \frac{\left(5 \cdot 10^{-3}\right)^2}{4} = 0.02244 \frac{\text{m}^3}{\text{s}}. \tag{6.28}$$

It can be approximated that a conventional compressor requires 0.1 electrical kW per Nm^3/h of compressed air produces. This ratio yields 8 kW of electrical power wasted, which means up to:

$$8\text{kW} \cdot 24 \frac{\text{h}}{\text{day}} \cdot 232 \frac{\text{day}}{\text{year}} = 44,544 \frac{\text{kWh}}{\text{year}}. \tag{6.29}$$

This is the potential saving for just one hole 5 mm wide. When multiplied for many uncontrolled losses, it makes advisable to think of a proper maintenance program.

6.10.4 *Boilers Fueled by Recovered Heat*

Boilers are common in a sanitary-ware factory. They are used to warm up water to flush the molds in a resin model-based factory. Besides, there are present for facilities as showers, water network, etc.

Thermal energy for boilers can be supplied by heated air recovered from the kiln in the same way as dryers and power plants. The rough estimation of savings would be as follows:

Thermal energy to warm up flushing water from 25 °C up to 45 °C:

$$0.18\frac{\mathrm{m}^3}{\mathrm{piece}}\cdot 500,000\frac{\mathrm{pieces}}{\mathrm{year}}\cdot 1,000\frac{\mathrm{kg}}{\mathrm{m}^3}\cdot 4.18\frac{\mathrm{kJ}}{\mathrm{kg}\cdot\mathrm{K}}\cdot(45-25)\mathrm{K}\cdot\frac{1\mathrm{h}}{3,600\mathrm{s}}$$
$$= 2.09\cdot 10^6\frac{\mathrm{kWh}}{\mathrm{year}}. \tag{6.30}$$

Thermal energy to warm up showers water from 25 °C up to 45 °C:

$$0.075\frac{\mathrm{m}^3}{\mathrm{shower}}\cdot 128\frac{\mathrm{showers}}{\mathrm{day}}\cdot 232\frac{\mathrm{days}}{\mathrm{year}}\cdot 1,000\frac{\mathrm{kg}}{\mathrm{m}^3}\cdot 4.18\frac{\mathrm{kJ}}{\mathrm{kg}\cdot\mathrm{K}}\cdot(45-25)\mathrm{K}\cdot\frac{1\mathrm{h}}{3,600\mathrm{s}}$$
$$= 51,720\frac{\mathrm{kWh}}{\mathrm{year}}. \tag{6.31}$$

6.10.5 *Actions on Pumping Systems*

For the pumping consumptions, all the electrical mentioned actions regarding motors and lighting are applicable here. Particularly the control of the frequency, since pumping allows large changes in the operation regime that result in large power reductions.

6.11 Other General Actions

There are some measures not directly linked to any section through the factory but to the global system that would produce great improvements on the global efficiency of the factory, such as.

6.11.1 Alternative Energies (Photovoltaic)

A practical and common current way of reducing electrical cost as well as emissions is by means of a photovoltaic plant. Photovoltaic power has been greatly developed during the last years from both, economic and technical points of view. Currently, it is of the utmost interest as energy efficiency investment. Legal regulations are different for any country or area, but in any case, the aim is to promote local generation as a common practice in the search of decarbonization objectives. During daylight hours, the energy generated is consumed and the excess is poured into the grid or the defect is taken from it at the purchase price. In the hours of shade, all the electricity is provided by public network. Known the curve of electrical demand as per the Fig. 6.11.

It could be sized a PV plant to supply a percentage of this demand during daylight time. The main question of this design would be this percentage. The most power generated by sun, the less electrical cost and the most plant price. Thus, there must be a detailed analysis of the electrical size of the plant in order to optimize the total cost of it. This kind of analysis is out of the scope of this book and there is uncountable literature regarding this subject (Table 6.12).

As an example, the following table shows the parameters of two PV plants sized to meet the maximum power demand, which is able to generate 36.21% of the annual demand and a greater one with coverage up to 43.37%.

Overlapping demand and generation curves, it can be estimated the appropriate size of the plant as shown in following graph (Fig. 6.12).

When the plant is designed as per the maximum demand power (1,728 kW), it can be seen that during the center of the day, the system is provided by the plant. The rest of the day and night, power has to be provided by the network. When the plant is enlarged in order to cover a wider range of hours (2,500 kW), it can be seen that generated power exceeds the requirement (the central peek). Thus, it is not worth to

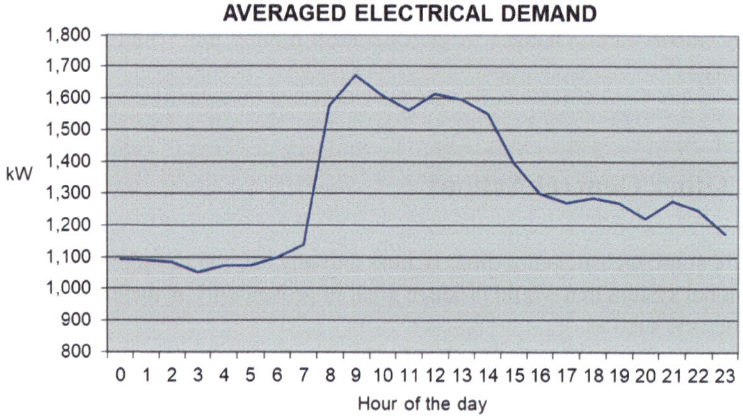

Fig. 6.11 Electrical demand of the factory (*Source* Data compiled by the authors)

Table 6.12 Operation parameters for two photovoltaic plants. (*Source* Data compiled by the authors)

Electrical demand			
Annual energy	kWh/year	7,669,258	
$P_{average}$	kW	1,377	
P_{max}	kW	1,763	
Photovoltaic Plant			
Total Power Plant	kW	1,763	2,500
Roof surface	m^2	23,100	23,100
Panels length	m	1.956	1.956
Panels width	m	0.992	0.992
Panels surface	m^2	1.940	1.940
Panel Power	W	300	300
Panel weigh	kg/panel	26	26
Structure weigh	kg/panel	29	29
Roof resistance	kg/m^2	37	37
Latitude	°	39.81	39.81
Projected length	m	1.503	1.503
projected surface	m^2	1.490	1.490
Efficiency		17.89%	17.89%
Occupancy rate of panels		37.91%	53.77%
Max number panels		15,499	15,499
Actual number panels		5,876	8,333
Averaged Radiation	$kWh/m^2 \cdot day$	4.60	4.60
Generated energy	kWh/year	2,777,388	3,939,010
Self-consumption	kWh/year	2,748,927	3,208,924
Coverage rate		35.84%	41.84%

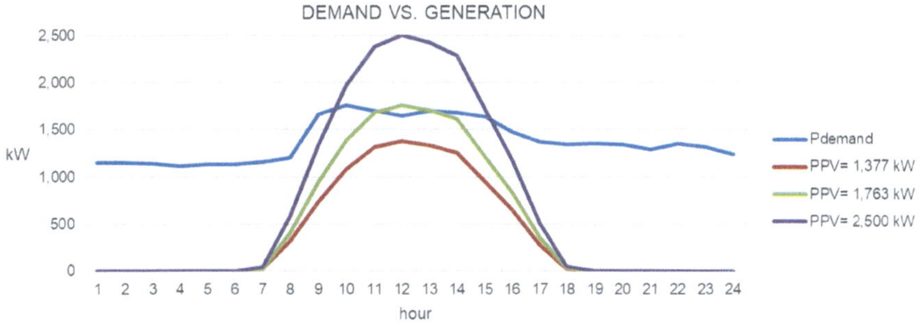

Fig. 6.12 Electrical demand versus different generation photovoltaic plants (*Source* Data compiled by the authors)

increase the power of the plant beyond the maximum demand unless, the excess may be sold to the local network.

6.11.2 Combined Heat and Power Plant

Though engine may be used to power the generator, usually gas turbine is preferred in this industrial sector since exhausting temperature is higher and more appropriated for the ceramic purposes. In a sanitary ceramic plant, the heat from the CHPP would supply heat to; dryers, heating plants, distillation or condensation of water for reuse. This heat is even, applicable to kiln combustion or auxiliary airs, since the high degree of oxygen contained in it.

This solution allows the kiln airs to be preheated to the temperature of the turbine exhaust gases, which are higher than the kiln chimney, so that the overall consumption of the kiln would decrease even more than preheating with chimney residual heat. On the other hand, the residual heat from the turbine allows feeding all the thermal receivers. However, some electricity might be still required and has to be supplied from local network. Figure 6.13 shows the configuration.

The issue is again the size of the GT. When too great (P_{max}), there is a need to either regulate power with lost efficiency or sell electrical surplus to the network. Besides, thermal power may result in exceeding requirements. Under such circumstances, the global efficiency may be decreased. On the contrary, when the GT is too small, there is not surplus of energy but it is not enough to cover the demand. In the example of the model, three GT have been tried to figure out the efficiency improvement as shown in Fig. 6.14.

The following table shows the potential savings for each.

Table 6.13 shows some facts, which could be deduced from the drawing. When the GT is designed as per the maximum of power, the total primary power is greater than a conventional system. This is due to the fact that much power is generated to be reverted to the network and leftover of heat is wasted. When averaged power is the size of the GT, efficiency is improved because the primary energy is less than the conventional system. Nevertheless, there is still a leftover of heat. In the case of designing GT as per the minimum demanded power, the efficiency improves further and shows that optimum efficiency is achieved when no heat leftover results.

In a conclusion, it has to be said that the design size of the GT has to be carried out considering that neither electrical nor thermal leftovers have to result.

6.11.3 Organic Rankine Cycles

ORCs are other potential sources of research since they can be run with medium temperature levels very suitable for kiln exhausting temperatures [16]. Electricity generation with low exergy grade heat through organic Rankine cycles [17] is

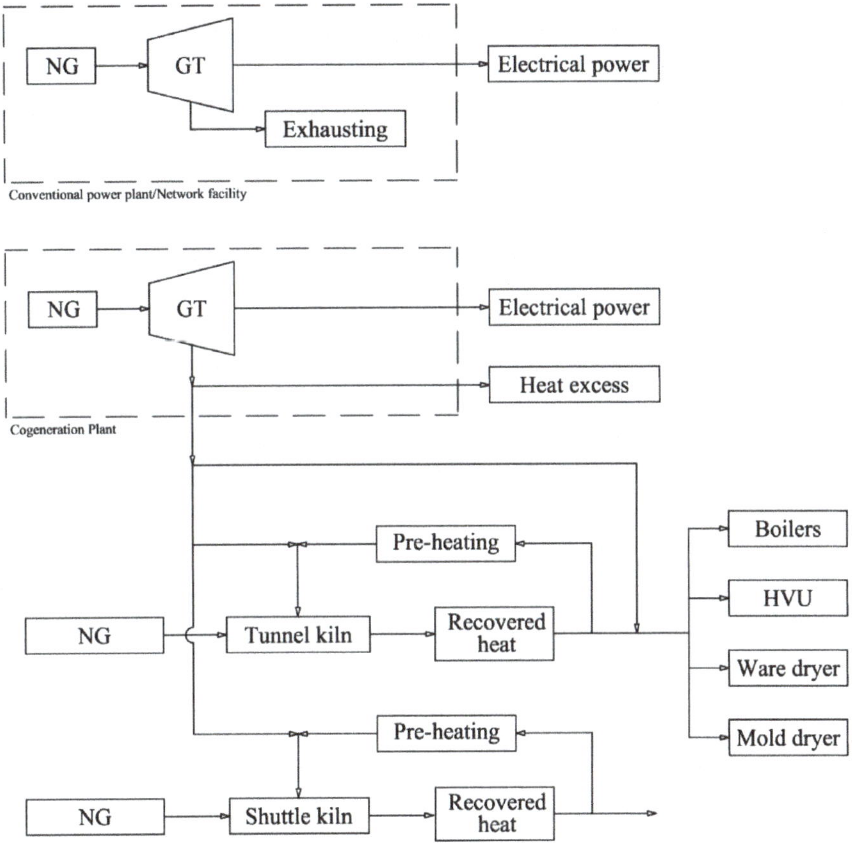

Fig. 6.13 Scheme of implemented combined heat and power plant in a factory

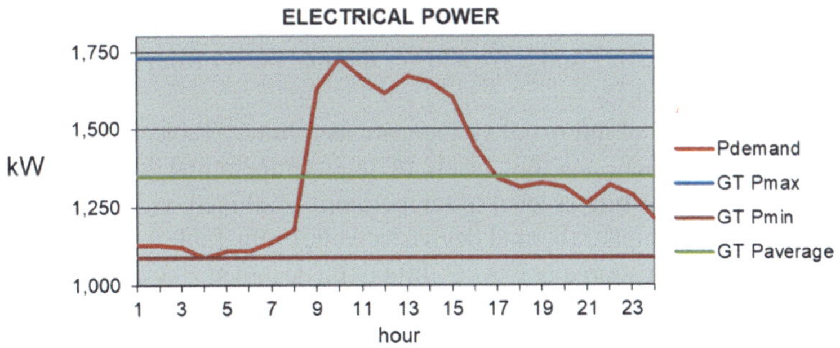

Fig. 6.14 Electrical demand versus different cogeneration plants

Table 6.13 Consumptions and savings for different cogeneration plant sizes

		Without GT	GT P_{max}	GT P_{min}	GT $P_{average}$
		kWh	kWh	kWh	kWh
Electricity	Generated power		9,814,963	6,189,345	7,669,258
	Consumed power	7,669,258	7,669,258	6,189,345	7,146,633
	Surplus of power		2,145,705	0	522,625
	Power from network=	7,669,258	0	1,479,913	522,625
Natural Gas	GT		32,716,543	20,631,149	25,564,193
	Tunnel kiln	14,524,898	14,524,898	14,524,898	14,524,898
	Shuttle kiln	4,275,871	4,275,871	4,275,871	4,275,871
	Molds dryer	2,004,792			
	Ware dryer	2,112,524			
	Heating plants	4,351,851			
	Boilers	581,856			
	Distiller				
	Absorption				
	Wrapping machine	240,000	240,000	240,000	240,000
	Canteen	18,560	18,560	18,560	18,560
	Recoverable heat from GT		21,592,918	13,616,558	16,872,367
	Recoverable heat from TK	12,106,582	12,106,582	12,106,582	12,106,582
	Recoverable heat from SK	1,626,931	1,626,931	1,626,931	1,626,931
	Residual heat		26,275,409	18,299,049	21,554,858
	Lack of recovery		0	0	0
	Natural gas from network	28,110,351	51,775,871	39,690,477	44,623,521
	Total Primary Power	53,674,544	51,775,871	44,623,521	46,365,604

produced by kiln exhausting and leave a lower thermal power from the condensation cycle as shown in Fig. 6.15.

Common capacities of industrial systems are as follows (Table 6.14).

As per the table, there are four levels of inlet temperature: low, medium and high. Besides, there is a higher level of temperature combined with a condensation cycle operating under higher thermal degree as well. From Table 5.25, the available temperature of the heat source is 288 °C. It may be designed a system to transform it into electricity through a heat exchanger air–water to supply any of the ORC's. Let us assume as an example, that a medium temperature is selected in the range of 150 °C. The system is shown in the Fig. 6.16.

From an energy balance, it can be calculated the output of the heat exchanger as the source of hot water for the ORC as follows:

$$m_{\text{exhausting}} \cdot (h_{\text{IN}} - h_{\text{OUT}}) = m_{\text{evaporator}} \cdot (h_{\text{IN}} - h_{\text{OUT}}). \tag{6.32}$$

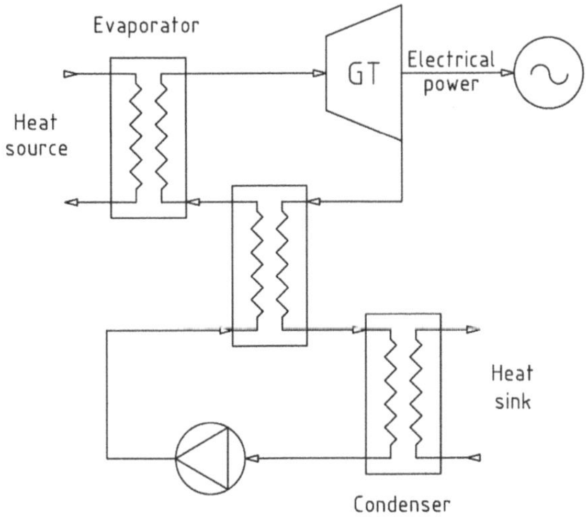

Fig. 6.15 Simplified scheme of an ORC

Table 6.14 Range of ORC parameters. (*Source* Data compiled by the authors)

		LT	MT	HT	HTC
Evaporator					
Fluid		Water	Water	Oil	Oil
Inlet T	°C	90–120	120–150	150–180	180–210
Outlet T	°C	80–110	110–140	110–140	120–150
Mass flow	m^3/h	17–78	17–78	13–56	13–44
Thermal power	kW	125–1,000	150–900	200–1,350	300–1,600
Condenser					
Fluid		Water	Water	Water	Water
Inlet T	°C	20–40	20–40	20–40	45–65
Outlet T	°C	30–50	30–50	30–50	60–80
Mass flow	m^3/h	14–63	14–63	17–77	60–80
Thermal power	kW	100–800	100–600	150–950	200–1200
Electrical power					
Gross power	kW	8–85	15–100	20–140	25–145
Net power	kW	8–80	13–90	18–120	20–115

With this flow of water and the ORC parameters, it can be calculated the performance of the ORC that summarized in Fig. 6.17:

It can be seen how kiln exhausting is able to supply 687 kW to produce 71.7 electrical kW and 434 thermal kW in the condenser water flow and still remain 733 kW of thermal power in the gas flow.

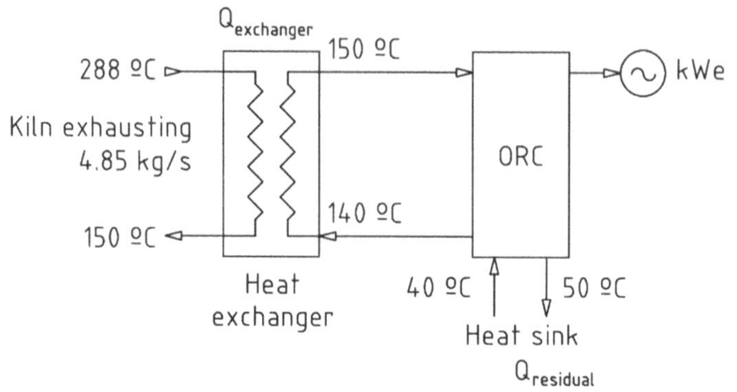

Fig. 6.16 Practical configuration of an ORC coupled to the kiln (*Source* Data compiled by the authors)

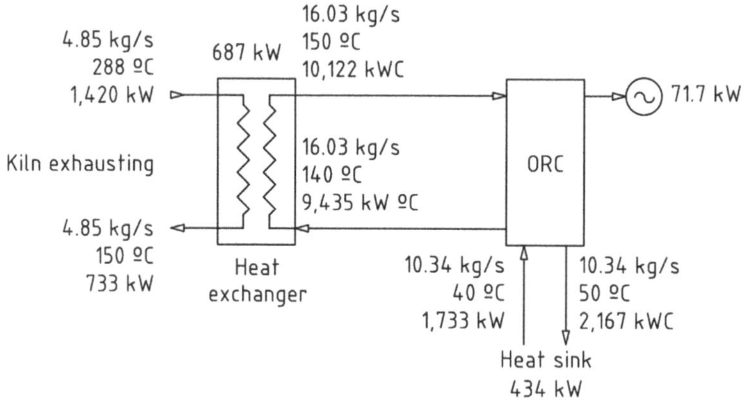

Fig. 6.17 Performance of the coupled ORC

6.12 Water-Saving Measures

As for water-saving measures, it can be said that they are not currently extended due to the relative low cost of water in factories, which generally does not justify reduction investments due to the long pay-back periods. In any case, as the cost of obtaining water increases, the viability of these investments will be reversed.

In Table 5.31, it can be seen the amount of water used and released by the factories. Depending on the origin of the water, it may have a different cost. It can be taken either from the local network or any available well. Usually, a conventional factory treats the wastewater before rejecting it back to the sewage. On the other hand, there is a fraction of the water that is released to the atmosphere as vapor through the exhausting of ventilation systems. Figure 6.18 shows, which processes require

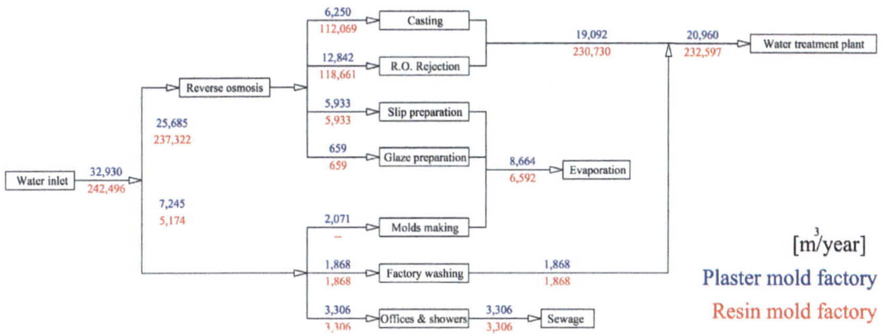

Fig. 6.18 Water distribution throughout the global process

treated water free of salts and what is the corresponding releasing via for every process.

A proper way to reuse these amounts of water is desirable from the environmental point of view beyond any economical aspect. Since there is residual heat from the kilns exhausting, it would be advisable to choose any system driven by thermal energy in order to optimize global efficiency. Distillation or reverse osmosis may be applied to residual liquid effluents and a condensation system to recover vaporized water operated with an absorption machine.

6.12.1 *Multi-effect Distillation*

Traditionally, reverse osmosis is the selected technology to treat process water within ceramic industry in order to eliminate dissolved salts.

From [18], it can be seen that 1 m^3 of freshwater requires 8.2 electrical kWh. On the contrary, multi-effect distillation requires only 3.8 electrical kWh plus 16.4 thermal kWh. Recoverable heat is available to supply this thermal demand [19]. Thus, 4.4 electrical kWh may be saved (54%). Besides, salinity of rejected water is the same in both cases and the production costs are at the same level. In a conclusion, it can be said that transferring water treatment from reverse osmosis to multi-effect distillation may be a good solution, especially for a resin mold factory where a large amount of water have to be treated free of dissolved salts to flush molds. Table 6.15 shows the requirements of water in both, plaster and resin mold factories and the way this water is released.

The liquid effluent from process wastewater equals 20,960 m^3/year for a plaster mold factory and 232,597 for a resin mold factory when a reverse osmosis plant is applied to treat process water. The efficiency of the plant is assumed to be 0.5 for this example.

In the event of transferring technology to MED, this water may be distilled to be reused. Besides, reverse osmosis capacity is reduced because of the reused water

Table 6.15 Water requirements in a conventional factory

Water distribution by R.O. technology (CF)					
		Plaster mold		Resin mold	
		m^3/year	m^3/year	m^3/year	m^3/year
Evaporated	Slip preparation	5,933	8,664	5,933	6,592
	Glaze preparation	659		659	
	Molds making	2,071		0	
Wastewater treatment	Molds flushing	6,250	20,960	112,069	232,597
	Factory washing	1,868		1,868	
	RO rejection	12,842		118,661	
Sewage	Offices &showers	3,306	3,306	3,306	3,306
	Total	32,930	32,930	242,496	242,496

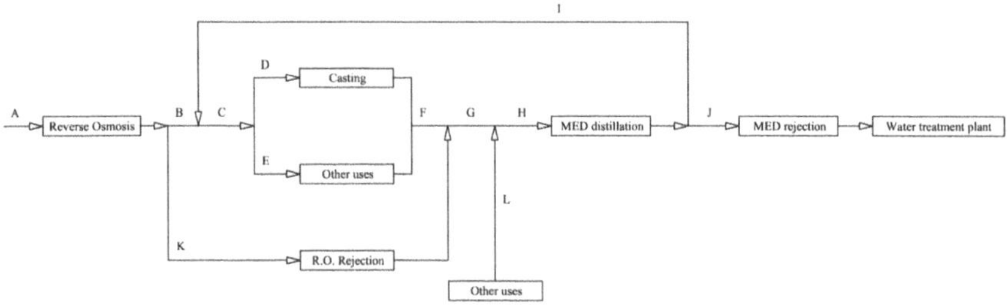

Fig. 6.19 Water reduction by MED distiller coupling

produced by distiller is already free of salts. This reduction of capacity implies as well a reduction of feeding and rejected water as shown in Fig. 6.19.

6.12.2 *Condensation via Absorption Machine*

As mentioned before, all the water contained in, slip, glaze and plaster is delivered to the atmosphere via evaporation through ventilation systems in dryers and hall heating. Evaporated water released by factories is 7,618 and 5,547 m^3/year, respectively, for plaster and resin mold factories. These amounts are not as representative as the liquid effluent, but they are worth being taken into account. Evaporated water can be condensed through a heat exchanger fed by the chilled flow from an absorption cycle as shown in Fig. 6.20.

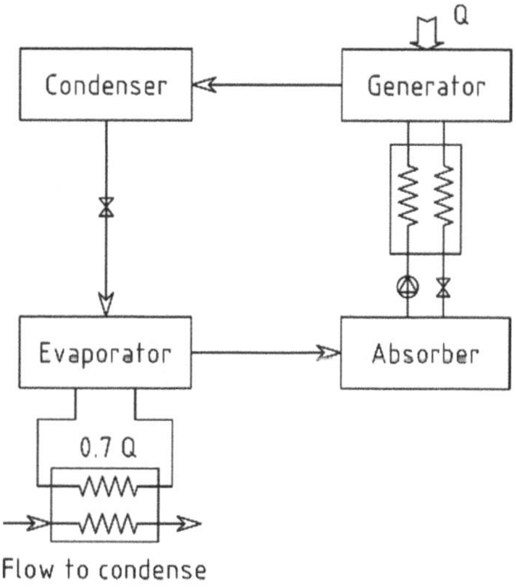

Fig. 6.20 Simplified scheme of an absorption machine

6.13 Optimization

All the saving measures seen up to now have been considered individually, that is, the calculated savings start from the conventional factory in which no measure had been previously implemented. However, there are some measures that are not compatible each other or at least interfere among them. Let us think, for instance, that a photovoltaic plant has to be designed. The power is different when it is designed for a standard factory where no electrical reduction measure has been carried out than when electrical reductions have been achieved. Under such circumstances, a combined power heat and power plant result smaller when a photovoltaic plant has been implemented before. Besides, the electrical demand curve changes drastically affected by photovoltaic electrical production. Regarding thermal savings, the approach is quite the same. When residual heat from kilns is taken into feeding airs, there is left less heat to be reused for dryers and heating. On the other hand, there is surplus of recoverable heat when cogeneration plant is implemented in the system. This action provides extra thermal power to apply into water savings via distillation and absorption.

To summarizing, there are multiple optimization variants that can be designed. The proper way to approach this design is to carry it out in consecutive stages, differentiating those independent measures from those that affect the global reduction.

Optimization criteria, usually, are not only related to energy, since there will always be an investment cost. It is quite common that the best energy solution is not the best economic one, so the problem becomes more complex considering also that the investment amortization depends strongly on the uncertain evolution of resource prices.

Table 6.16 Savings in the first stage

Savings by general measures	kWh/year	
Electrical measures	Plaster molds	Resin molds
Savings by LED lighting	79,680	63,410
Savings by motor efficiency	201,922	141,303
Savings by lighting regulation	30,886	24,639
Savings by frequency converters	1,609,294	1,273,461
Total Electrical	1,921,782	1,502,813
Thermal Measures		
Thermal insulation	495,890	189,594
Regulation of fresh air intake	487,363	144,142
Load lightening by SiC	2,917,269	2,917,269
Heat recovery	569,600	569,600
Total thermal	4,470,122	3,820,606

In this book, as an example, the optimization process has been carried out in three sequential stages. It starts from the consumption of the conventional factory without any saving measure whose detailed consumptions were shown in Table 5.32.

6.13.1 First Stage

In this first stage, they have been considered those measures which do not interfere with each other. They can be summarized in Table 6.16.

6.13.2 Second Stage

In this stage, there is a range of measures that cannot be applied as a whole. In Table 6.17, it can be seen the potential savings for any of these actions individually. Negative values mean an increase in consumption instead of saving.

In order to arrange the selection of measures and design it implementation, a plan has to be made to optimize savings.

Heat recovery from kilns may not be able to supply the rest of heat consumers. In a standard kiln where neither induced draught nor air preheating has been applied, the annual recoverable amount of heat is 12,106,582 kWh. This heat is enough to supply; dryers, heating units and boilers. However, when part of this heat is derived to pre-heat the primary air of the kilns, the resultant recoverable heat decreases to 5,064,305 kWh, which is not able to cover the thermal demand. For the purpose of optimizing the total consumption of resources, the sequential procedure to achieve the minimum is as follows:

Table 6.17 Savings in the second stage

Section	Type	Plaster molds			Resin molds		
		Electricity	Gas	Water	Electricity	Gas	Water
		kWh/year	kWh/year	m^3/year	kWh/year	kWh/year	m^3/year
Modeling and molds	Heat recovery for mold dryers		2,004,792				
Heating and ventilation	Heat recovery for heating units					1,118,246	
Drying	Heat recovery for ware dryers		4,351,851			609,956	
Kilns	Induced draught for tunnel kiln		2,112,526			2,112,526	
Kilns	Primary air pre heating for tunnel kiln		581,856			581,856	
Kilns	Primary air pre heating for shuttle kiln		3,086,298			3,086,298	
Facilities	Heat recovery for boilers		3,594,441			3,594,441	
Global	Condensation of vapor by absorption		1,931,509			1,931,509	
Global	MED distillation	−27,840	-4,277,379	4,412	−27,840	−2,291,877	2,364
Global	Photovoltaic plant	−96,016	-1,475,110	12,842	−887,161	−13,629,616	118,661
Global	Cogeneration plant	2,748,927			2,250,072		

- Use residual heat to optimize kiln consumption: induced draught and air pre-heating.
- The rest of heat is recovered toward dryers, heating units and boilers.
- A MED distiller is applied to reduce water requirements. Electrical and thermal demand increases to operate it.
- An absorption machine is applied to condense evaporated water. Electrical and thermal demand increases to operate it.
- A photovoltaic plant is applied to reduce electrical consumption. It has to be sized after decreasing the electrical savings already achieved.

- A cogeneration plant is designed to cover the electrical demand during nighttime. In order to optimize thermal consumption, the criteria to size the plant must be to follow the thermal demand rather than electrical because electrical leftovers might be download to the network. However, it is a waste of a thermal leftover.

6.14 Global Assessment

As stated in Chap. 1, in 2014, the global sanitary-ware production reached 349.3 million pieces where 59 million were produced in Europe. With the assessed factory models, energy and water consumption can be deduced globally for the sanitary ware industry. In Table 6.18, the figures for these productions are shown.

Table 6.18 gives a clear picture to understand the great opportunity for energy saving within the sanitary-ware industry. Therefore, there is a wide field to implement this technology worldwide providing a more sustainable and profitable business from sanitary-ware manufacturing.

In this growing world, there is a real need to develop these techniques in order to improve the efficiency further [20]. As per WHO, 2.3 billion people over 7.3 billion had not any basic sanitation facilities such as toilets or latrines in 2015. In 2013, the UN Deputy Secretary General issued a call to action on sanitation that included the elimination of open defecation by 2025. Achieving universal access to a basic drinking water source appears within reach, but universal access to basic sanitation will require additional efforts [21]. This means that some billions of sanitary-ware articles are to be manufactured within short term, thus better factories have to be designed with the most advanced and efficient technologies to optimize the environmental impact in those developing countries where sanitation efforts are to be made.

Different energy and resource efficiency recycling methods as well as reuse resources strategies obtainable by proper implementation of manufacturing processes are suggested, so as potential improvements as a tool to optimize resources. Hence, the applicability of the tool by modeling manufacturing processes is a suitable means to design the most convenient manufacturing line for any background as a suitable tool to support a lean manufacturing philosophy.

Industrial and economical decisions are made by taking into account resource consumption, environmental impacts of them and finally, the way both concepts are

Table 6.18 Consumption for global sanitary ware production

	Production	Plaster Mold			Resin Mold		
		Electricity	Gas	Water	Electricity	Gas	Water
	Pieces/year	GWh/year	GWh/year	Hm^3/year	GWh/year	GWh/year	Hm^3/year
EU	41.5	637	2,333	2.73	521	1,949	20.13
World	349.3	5,358	19,638	23.00	4,385	16,404	169.41

linked for a more efficient and sustainable industry and a cleaner environment around it. This paper aims to provide a means to support such an effort to be developed.

References

1. Boharb A et al (2016) Auditing and analysis of energy consumption of an industrial site in Morocco. Energy 101:332–342. https://doi.org/10.1016/j.energy.2016.02.035
2. Uhlmann E, Reinkober S, Hollerbach T (2016) Energy efficient usage of industrial robots for machining processes. Proc CIRP 48:206–211. https://doi.org/10.1016/j.procir.2016.03.241
3. Vittorini D, Cipollone R (2016) Energy saving potential in existing industrial compressors. Energy 102:502–515. https://doi.org/10.1016/j.energy.2016.02.115
4. Yin Q, Du W-J, Ji X-L, Cheng L (2016) Optimization design and economic analyses of heat recovery exchangers on rotary kilns. Appl Energy 180:743–756. https://doi.org/10.1016/j.apenergy.2016.07.042
5. Utlu Z, Hepbaşlı A (2014) Exergoeconomic analysis of energy utilization of drying process in a ceramic production. Appl Therm Eng 70(1):748–762. https://doi.org/10.1016/j.applthermaleng.2014.05.070
6. Merlin K, Soto J, Delaunay D, Traonvouez L (2016) Industrial waste heat recovery using an enhanced conductivity latent heat thermal energy storage. Appl Energy 183:491–503. https://doi.org/10.1016/j.apenergy.2016.09.007
7. Hasan M, Zhang M, Wu W, Langrish TAG (2016) Discounted cash flow analysis of greenhouse-type solar kilns. Renew Energy 95:404–412. https://doi.org/10.1016/j.renene.2016.04.050
8. Chiaroni D, Chiesa M, Chiesa V, Franzò S, Frattini F, Toletti G (2016) Introducing a new perspective for the economic evaluation of industrial energy efficiency technologies: An empirical analysis in Italy. Sustain Energy Technol Assess 15:1–10. https://doi.org/10.1016/j.seta.2016.02.004
9. Li C, Goswami DY, Shapiro A, Stefanakos EK, Demirkaya G (2012) A new combined power and desalination system driven by low grade heat for concentrated brine. Energy 46(1):582–595. https://doi.org/10.1016/j.energy.2012.07.050
10. Colmenar-Santos A, Campíñez-Romero S, Pérez-Molina C, Mur-Pérez F (2016) An assessment of photovoltaic potential in shopping centres. Sol Energy 135:662–673. https://doi.org/10.1016/j.solener.2016.06.049
11. Borge-Diez D, Colmenar-Santos A, Pérez-Molina C, Castro-Gil M (2012) Experimental validation of a fully solar-driven triple-state absorption system in small residential buildings. Energy Build 55:227–237. https://doi.org/10.1016/j.enbuild.2012.09.040
12. Schreiber H, Graf S, Lanzerath F, Bardow A (2015) Adsorption thermal energy storage for cogeneration in industrial batch processes: experiment, dynamic modeling and system analysis. Appl Therm Eng 89:485–493. https://doi.org/10.1016/j.applthermaleng.2015.06.016
13. Wang X, El-Farra NH, Palazoglu A (2017) Optimal scheduling of demand responsive industrial production with hybrid renewable energy systems. Renew Energy 100:53–64. https://doi.org/10.1016/j.renene.2016.05.051
14. Colmenar-Santos A, Rosales-Asensio E, Borge-Diez D, Blanes-Peiró J-J (2016) District heating and cogeneration in the EU-28: current situation, potential and proposed energy strategy for its generalisation. Renew Sustain Energy Rev 62:621–639. https://doi.org/10.1016/j.rser.2016.05.004
15. Sipilä K (2016) 3 - cogeneration, biomass, waste to energy and industrial waste heat for district heating A2 - Wiltshire, Robin. Advanced District Heating and Cooling (DHC) systems. Woodhead Publishing, Oxford, pp 45–73
16. Ustaoglu A, Alptekin M, Akay ME (2017) Thermal and exergetic approach to wet type rotary kiln process and evaluation of waste heat powered ORC (Organic Rankine Cycle). Appl Therm Eng 112:281–295. https://doi.org/10.1016/j.applthermaleng.2016.10.053

17. Organic Rankine Cycle - an overview|ScienceDirect Topics (2020). https://www.sciencedirect.com/topics/engineering/organic-rankine-cycle. Accessed 27 Oct 2020
18. Youssef PG, AL-Dadah RK, Mahmoud SM (2014) Comparative analysis of desalination technologies. Energy Proc 61:2604–2607. https://doi.org/10.1016/j.egypro.2014.12.258
19. Ge Z, Yang C, Liu Y, Du X, Yang L, Yang Y (2014) Analysis of plate multi-effect distillation system coupled with thermal power generating unit. Appl Therm Eng 67(1):35–42. https://doi.org/10.1016/j.applthermaleng.2014.02.072
20. Berg SV (2015) Energy efficiency in developing countries: Roles for sector regulators. Energy Sustain Dev 29:72–79. https://doi.org/10.1016/j.esd.2015.10.002
21. WHO Sanitation WHO news-room fact sheets, June 19. https://www.who.int/news-room/fact-sheets/detail/sanitation

Chapter 7
Proposals Calculation

7.1 Electrical Savings

Consumptions through the factory have been computed in Chap. 5. By Table 5.32, they can be derived the most representative of them in which actuate by taking corrective measures in order to reduce single consumptions. This table provides as well the location where the most consumptions are concentrated. Hence, by means of the enumerated actions defined within Chap. 6, they can be computed generic electrical savings, which are described in the following sections.

7.1.1 Efficiency of the Motors

Table 7.1 shows the savings that may be achieved by increasing electrical motor efficiency as described in Sect. 6.1.1.

7.1.2 Frequency Variation of Electrical Motors

Assuming a speed reduction of 10% of the sized and based on Tables 5.1–5.13, the savings for the whole factory aggregated by section result as follows (Table 7.2).

7.1.3 Substitution of Conventional by LED Lighting

According to Table 5.14 and Sect. 6.1.3, Table 7.3 shows the potential savings from LED technology instead of conventional lighting:

C. Cuviella-Suárez et al., *Water and Energy Use in Sanitary-ware Manufacturing*, Green Energy and Technology, https://doi.org/10.1007/978-3-030-72491-7_7

Table 7.1 Savings derived from increasing efficiency of the electrical motors, from IE2 to IE3

Facilities		Savings by motor efficiency													
		Plaster mold factory							Resin mold factory						
		Number of equips	Motors per equip	Motor power	e_{IE3}	e_{IE2}	*De*	Savings	Number of equips	Motors per equip	Motor power	e_{IE3}	e_{IE2}	*De*	Savings
				kW	%	%	%	kWh/year			kW	%	%	%	kWh/year
DB01	Raw materials							9,317							9,317
DB01-A	Mills	2	1	110.0	0.96	0.94	0.02	7,656	2	1	110.0	0.96	0.94	0.02	7,656
DB01-B	Belts	7	1	5.0	0.89	0.87	0.03	1,277	7	1	5.0	0.89	0.87	0.03	1,277
DB01-C	Pneumatic pumps	3							3						
DB01-D	Hoppers	6	1	2.5	0.87	0.85	0.02	384	6	1	2.5	0.87	0.85	0.02	384
DB02	Slip Preparation							22,661							22,661
DB02-A	Blungers	3	1	125.0	0.96	0.94	0.02	8,700	3	1	125.0	0.96	0.94	0.02	8,700
DB02-B	Sieves	6	1	1.5	0.85	0.83	0.03	344	6	1	1.5	0.85	0.83	0.03	344
DB02-C	Peristaltic pumps	3	1	15.0	0.92	0.90	0.02	908	3	1	15.0	0.92	0.90	0.02	908
DB02-D	Slip tanks	7	1	15.0	0.92	0.90	0.02	12,710	7	1	15.0	0.92	0.90	0.02	12,710
DB03	Glaze Preparation							7,222							7,222
DB03-A	Mills	2	1	25.0	0.94	0.92	0.02	1,340	2	1	25.0	0.94	0.92	0.02	1,340
DB03-B	Belts	2	1	5.0	0.89	0.87	0.03	261	2	1	5.0	0.89	0.87	0.03	261
DB03-C	Blungers	2	1	30.0	0.94	0.92	0.02	1,072	2	1	30.0	0.94	0.92	0.02	1,072

(continued)

Table 7.1 (continued)

Facilities		Savings by motor efficiency													
		Plaster mold factory							Resin mold factory						
		Number of equips	Motors per equip	Motor power	e_{IE3}	e_{IE2}	*De*	Savings	Number of equips	Motors per equip	Motor power	e_{IE3}	e_{IE2}	*De*	Savings
				kW	%	%	%	kWh/year			kW	%	%	%	kWh/year
DB03-D	Sieves	3	1	1.5	0.85	0.83	0.03	172	3	1	1.5	0.85	0.83	0.03	172
DB03-E	Pneumatic pumps	2							2						
DB03-F	Glaze tanks	19	1	3.0	0.87	0.85	0.02	4,378	19	1	3.0	0.87	0.85	0.02	4,378
DB04	Modeling and Molds							15,405							677
DB04-A	Molds dryers	3	7	5.0	0.89	0.87	0.03	14,780	1	7	0.7	0.89	0.87	0.03	235
DB04-B	Cranes	10	1	1.8	0.85	0.83	0.03	382	1	1	1.8	0.85	0.83	0.03	38
DB04-C	Plaster preparation machine	1	2	2.5	0.86	0.84	0.03	243	1	2	2.5	0.86	0.84	0.03	405
DB04-D	Pneumatic pumps	3							2						
DB05	Casting							71,606							27,591
DB05-A	Benches	30	40	0.5	0.80	0.76	0.05	66,816	13	40	0.5	0.80	0.76	0.05	22,801

(continued)

Table 7.1 (continued)

Facilities		Savings by motor efficiency													
		Plaster mold factory							Resin mold factory						
		Number of equips	Motors per equip	Motor power	e_{IE3}	e_{IE2}	*De*	Savings	Number of equips	Motors per equip	Motor power	e_{IE3}	e_{IE2}	*De*	Savings
				kW	%	%	%	kWh/year			kW	%	%	%	kWh/year
DB05-C	Ponds	2	2	2.5	0.86	0.84	0.03	324	2	2	2.5	0.86	0.84	0.03	324
DB05-D	Tanks	2	1	25.0	0.94	0.92	0.02	4,466	2	1	25.0	0.94	0.92	0.02	4,466
DB05-E	Pneumatic pumps	4													
DB06	Heating and Ventilation							3,197							3,197
DB06-A	Heating and ventilation unit	1	2	25.0	0.94	0.92	0.02	2,233	1	2	25.0	0.94	0.92	0.02	2,233
DB06-B	Extraction fans	6	1	5.5	0.89	0.87	0.02	963	6	1	5.5	0.89	0.87	0.02	963
DB07	Drying							28,823							28,823
DB07-A	Ware dryers	2	15	3.7	0.77	0.73	0.05	28,823	2	15	3.7	0.77	0.73	0.05	28,823
DB08	Glazing							4,857							4,857
DB08-A	Carrousels	2	10	4.5	0.89	0.87	0.02	3,397	2	10	4.5	0.89	0.87	0.02	3,397

(continued)

Table 7.1 (continued)

Facilities		Savings by motor efficiency													
		Plaster mold factory							Resin mold factory						
		Number of equips	Motors per equip	Motor power	e_{IE3}	e_{IE2}	*De*	Savings	Number of equips	Motors per equip	Motor power	e_{IE3}	e_{IE2}	*De*	Savings
				kW	%	%	%	kWh/year			kW	%	%	%	kWh/year
DB08-B	Glazing boots	8	1	5.0	0.89	0.87	0.03	1,460	8	1	5.0	0.89	0.87	0.03	1,460
DB09	Kilns							14,910							14,910
DB09-A	Tunnel kiln	1	5	25.0	0.94	0.92	0.02	8,663	1	5	25.0	0.94	0.92	0.02	8,663
DB09-B	Shuttle kiln	1	5	20.0	0.94	0.92	0.02	5,775	1	5	20.0	0.94	0.92	0.02	5,775
DB09-C	Second fire boot	2	1	5.0	0.89	0.87	0.03	472	2	1	5.0	0.89	0.87	0.03	472
DB10	Sorting							3,363							3,363
DB10-A	Sorting lines	4	7	0.7	0.82	0.78	0.05	2,371	4	7	0.7	0.82	0.78	0.05	2,371
DB10-B	Wrapping machine	1	3	3.3	0.88	0.85	0.03	480	1	3	3.3	0.88	0.85	0.03	480
DB10-C	Lifting tables	4	1	4.8	0.89	0.87	0.02	512	4	1	4.8	0.89	0.87	0.02	512
DB11	Offices							971							971
DB11-A	Office desk	15							15						
DB11-B	IT central servers	1							1						

(continued)

Table 7.1 (continued)

Facilities		Savings by motor efficiency													
		Plaster mold factory							Resin mold factory						
		Number of equips	Motors per equip	Motor power	e_{IE3}	e_{IE2}	*De*	Savings	Number of equips	Motors per equip	Motor power	e_{IE3}	e_{IE2}	*De*	Savings
				kW	%	%	%	kWh/year			kW	%	%	%	kWh/year
DB11-C	Conditioning air	10	2	2.3	0.86	0.84	0.03	971	10	2	2.3	0.86	0.84	0.03	971
DB11-D	Other consumers	1							1						
DB12	Ware House							0							0
DB12-A	Office	1							1						
DB12-B	Batteries charging	5							5						
DB13	Facilities							19,589							17,713
DB13-A	Compressors	2	1	160.0	0.96	0.95	0.01	7,463	2	1	160.0	0.96	0.95	0.01	7,463
DB13-B	Boilers	1	2	2.5	0.86	0.84	0.03	486	1	2	2.5	0.86	0.84	0.03	486
DB13-C	Maintenance work shop	1	15	3.0	0.88	0.85	0.03	1,193	1	15	3.0	0.88	0.85	0.03	1,193

(continued)

Table 7.1 (continued)

Facilities		Savings by motor efficiency													
		Plaster mold factory							Resin mold factory						
		Number of equips	Motors per equip	Motor power	e_{IE3}	e_{IE2}	*De*	Savings	Number of equips	Motors per equip	Motor power	e_{IE3}	e_{IE2}	*De*	Savings
				kW	%	%	%	kWh/year			kW	%	%	%	kWh/year
DB13-D	Losses hopper and crushing machine	1	10	5.0	0.89	0.87	0.02	1,043	1	10	5.0	0.89	0.87	0.02	1,043
DB13-E	Pumps and reverse osmosis plant	10	1	5.0	0.89	0.87	0.02	3,754	1	1	25.0	0.89	0.87	0.02	1,877
DB13-F	Wastewater plant	1	10	7.5	0.90	0.87	0.03	5,568	1	10	7.5	0.90	0.87	0.03	5,568
DB13-G	Security room	1							1						
DB13-H	Waste treatment	1	1	5.0	0.89	0.87	0.02	83	1	1	5.0	0.89	0.87	0.02	83
DB13-I	Lunch room	1							1						
DB13-J	Canteen	1							1						
DB13-K	External lighting	1							1						

(continued)

Table 7.1 (continued)

Facilities		Savings by motor efficiency													
		Plaster mold factory							Resin mold factory						
		Number of equips	Motors per equip	Motor power	e_{IE3}	e_{IE2}	*De*	Savings	Number of equips	Motors per equip	Motor power	e_{IE3}	e_{IE2}	*De*	Savings
				kW	%	%	%	kWh/year			kW	%	%	%	kWh/year
DB13-L	Plugs circuit	1							1						
DB13-M	Watering	1							1						
DB13-N	Electrical station	1							1						
DB13-O	Gas station	1							1						
DB13-P	Generator	1							1						
DB13-Q	Water tank reservoir	1							1						
	Total Motor Efficiency	214						201,922	174						141,303

Table 7.2 Savings derived from variations of speed by means of frequency converters. (*Source* Data compiled by the authors)

Facilities		Savings by frequency converter									
		Plaster mold factory					Resin mold factory				
		Number of equips	Converters per equip	Motor power	*Dn*	Savings	Number of equips	Converters per equip	Motor power	*Dn*	Savings
				kW	%	kWh/year			kW	%	kWh/year
DB01	Raw materials					36,749					36,749
DB01-A	Mills	2	1	110.0	10.00	36,749	2	1	110.0	10.00	36,749
DB01-B	Belts	7					7				
DB01-C	Pneumatic pumps	3					3				
DB01-D	Hoppers	6					6				
DB02	Slip Preparation					282,924					282,924
DB02-A	Blungers	3	1	125.0	10.00	113,170	3	1	125.0	10.00	113,170
DB02-B	Sieves	6					6				
DB02-C	Peristaltic pumps	3	1	15.0	10.00	11,317	3	1	15.0	10.00	11,317
DB02-D	Slip tanks	7	1	15.0	10.00	158,437	7	1	15.0	10.00	158,437
DB03	Glaze Preparation					92,346					92,346
DB03-A	Mills	2	1	25.0	10.00	22,634	2	1	25.0	10.00	22,634
DB03-B	Belts	2					2				
DB03-C	Blungers	2	1	30.0	10.00	18,107	2	1	30.0	10.00	18,107
DB03-D	Sieves	3					3				

(continued)

Table 7.2 (continued)

Facilities		Savings by frequency converter									
		Plaster mold factory					Resin mold factory				
		Number of equips	Converters per equip	Motor power	*Dn*	Savings	Number of equips	Converters per equip	Motor power	*Dn*	Savings
				kW	%	kWh/year			kW	%	kWh/year
DB03-E	Pneumatic pumps	2					2				
DB03-F	Glaze tanks	19	1	3.0	10.00	51,605	19	1	3.0	10.00	51,605
DB04	Modeling and Molds					0					0
DB04-A	Molds dryers	3					1				
DB04-B	Cranes	10					1				
DB04-C	Plaster preparation machine	1					1				
DB04-D	Pneumatic pumps	3					2				
DB05	Casting					391,664					78,464
DB05-A	Benches	30	45	0.125	50.00	313,200	13				
DB05-C	Ponds	2	2	2.5	10.00	3,018	2	2	2.5	10.00	3,018
DB05-D	Tanks	2	1	25.0	10.00	75,446	2	1	25.0	10.00	75,446

(continued)

Table 7.2 (continued)

Facilities		Savings by frequency converter									
		Plaster mold factory					Resin mold factory				
		Number of equips	Converters per equip	Motor power	*Dn*	Savings	Number of equips	Converters per equip	Motor power	*Dn*	Savings
				kW	%	kWh/year			kW	%	kWh/year
DB05-E	Pneumatic pumps	4					2				
DB06	Heating and Ventilation					37,723					37,723
DB06-A	Heating and ventilation unit	1	2	25.0	10.00	37,723	1	2	25.0	10.00	37,723
DB06-B	Extraction fans	6					6				
DB07	Drying					149,384					149,384
DB07-A	Ware dryers	2	15	4	10.00	149,384	2	15	4	10.00	149,384
DB08	GLAZING					0					0
DB08-A	Carrousels	2					2				
DB08-B	Glazing boots	8					8				
DB09	Kilns										243,900
DB09-A	Tunnel kiln	1	5	25.0	10.00	146,340	1	5	25.0	10.00	146,340
DB09-B	Shuttle kiln	1	5	20.0	10.00	97,560	1	5	20.0	10.00	97,560

(continued)

Table 7.2 (continued)

Facilities		Savings by frequency converter									
		Plaster mold factory					Resin mold factory				
		Number of equips	Converters per equip	Motor power	*Dn*	Savings	Number of equips	Converters per equip	Motor power	*Dn*	Savings
				kW	%	kWh/year			kW	%	kWh/year
DB09-C	Second fire boot	2					2				
DB10	Sorting					0					0
DB10-A	Sorting lines	4					4				
DB10-B	Wrapping machine	1					1				
DB10-C	Lifting tables	4					4				
DB11	Offices					9,054					9,054
DB11-A	Office desk	15					15				
DB11-B	IT central servers	1					1				
DB11-C	Conditioning air	10	2	2.3	10.00	9,054	10	2	2.3	10.00	9,054
DB11-D	Other consumers	1					1				
DB12	Ware House					0					0
DB12-A	Office	1					1				

(continued)

Table 7.2 (continued)

Facilities		Savings by frequency converter									
		Plaster mold factory					Resin mold factory				
		Number of equips	Converters per equip	Motor power	*Dn*	Savings	Number of equips	Converters per equip	Motor power	*Dn*	Savings
				kW	%	kWh/year			kW	%	kWh/year
DB12-B	Batteries charging	5					5				
DB13	Facilities					365,550					342,916
DB13-A	Compressors	2	1	160.0	15.00	275,015	2	1	150.0	15.00	275,015
DB13-B	Boilers	1					1				
DB13-C	Maintenance work shop	1					1				
DB13-D	Losses hopper and crushing machine	1					1				
DB13-E	Pumps and reverse osmosis plant	10	1	5.0	10.00	45,268	1	1	25.0	10.00	22,634
DB13-F	Wastewater plant	1	10	7.5	10.00	45,268	1	10	7.5	10.00	45,268
DB13-G	Security room	1					1				

(continued)

Table 7.2 (continued)

Facilities		Savings by frequency converter									
		Plaster mold factory					Resin mold factory				
		Number of equips	Converters per equip	Motor power	*Dn*	Savings	Number of equips	Converters per equip	Motor power	*Dn*	Savings
				kW	%	kWh/year			kW	%	kWh/year
DB13-H	Waste treatment	1					1				
DB13-I	Lunch room	1					1				
DB13-J	Canteen	1					1				
DB13-K	External lighting	1					1				
DB13-L	Plugs circuit	1					1				
DB13-M	Watering	1					1				
DB13-N	Electrical station	1					1				
DB13-O	Gas station	1				0	1				0
DB13-P	Generator	1				0	1				0
DB13-Q	Water tank reservoir	1				0	1				0
	Total frequency converters					1,609,294					1,273,461

Table 7.3 Electrical savings by substituting conventional lighting by LED

Lighting levels	Savings by LED lighting					
	Plaster mold factory			Resin mold factory		
	$P_{CONVENTtIONAL}$	P_{LED}	Savings	$P_{CONVENTtIONAL}$	P_{LED}	Savings
	kW	kW	kWh/yr	kW	kW	kWh/yr
Raw materials	3.38	2.41	1,790	3.38	2.41	1,790
Slip preparation	2.50	1.79	1,326	2.50	1.79	1,326
Glaze preparation	3.52	2.34	2,175	3.52	2.34	2,175
Modeling and molds	17.58	11.72	10,875	17.58	11.72	10,875
Casting	40.63	27.08	25,133	13.54	9.03	8,378
Heating and ventilation	0.78	0.52	483	0.78	0.52	483
Drying	7.81	5.21	4,833	7.81	5.21	4,833
Glazing	19.53	13.02	12,083	19.53	13.02	12,083
Kilns	10.16	6.77	6,283	10.16	6.77	6,283
Sorting	7.81	5.21	4,833	7.81	5.21	4,833
Offices	3.91	2.60	2,417	3.91	2.60	2,417
Ware house	3.91	2.60	2,417	3.91	2.60	2,417
Facilities*	9.02	6.31	5,031	9.83	6.86	5,517
Total LED lighting=	130.52	87.58	79,680	104.24	70.08	63,410
*Facilities split up						
Compressors	0.31	0.21	193	0.31	0.21	193
Boilers	0.08	0.05	48	0.08	0.05	48
Maintenance work shop	1.17	0.78	725	1.17	0.78	725
Losses hopper and crushing machine	0.63	0.45	331	0.63	0.45	331
Pumps and reverse osmosis plant	0.31	0.21	193	0.94	0.63	580
Wastewater plant	0.44	0.31	232	0.63	0.45	331
Security room	0.09	0.06	58	0.09	0.06	58
Waste treatment	0.02	0.01	10	0.02	0.01	10
Lunch room	0.16	0.10	97	0.16	0.10	97

(continued)

Table 7.3 (continued)

Lighting levels	Savings by LED lighting					
	Plaster mold factory			Resin mold factory		
	$P_{CONVENTIONAL}$	P_{LED}	Savings	$P_{CONVENTIONAL}$	P_{LED}	Savings
	kW	kW	kWh/yr	kW	kW	kWh/yr
Canteen	0.06	0.04	39	0.06	0.04	39
External lighting	4.69	3.35	2,486	4.69	3.35	2,486
Plugs circuit						
Watering						
Electrical station	0.47	0.31	290	0.47	0.31	290
Gas station	0.13	0.09	66	0.13	0.09	66
Generator	0.16	0.10	97	0.16	0.10	97
Water tank reservoir	0.31	0.22	166	0.31	0.22	166

7.1.4 Lighting Regulation by Control of Presence and External Lighting Level

According to the model of lighting regulation in Sect. 6.1.4, the savings for the whole factory aggregated by section result as follows (Table 7.4).

7.1.5 Summary of Electrical Generalized Actions

Up to this point, the called generalized actions have been applied and computed for common equipment in order to reduce electrical consumption. Table 7.5 summarizes their results.

7.2 Non-exclusive Thermal Savings

Thermal saving possibilities analyzed in Chap. 6 can be exclusive each other. Thus, they are going to be split in two parts. Those which can be applied regardless other measures (non-exclusive) and the rest of them which are depending on each other. Heat recovery, as an example, may not supply all consumers of heat. Under such circumstances, it can be applied to some consumers up to the point that there is left nothing. At this point, this measure is not applicable for distilling for example. These are going to be called exclusive ones.

Table 7.4 Electrical savings by lighting regulation. (*Source* Data compiled by the authors)

Savings by lighting regulation				
Lighting levels	f_{presence}	f_{exterior}	PMF	RMF
			Savings	Savings
	–	–	kWh/yr	kWh/yr
Raw materials	0.10	0.10	850	850
Slip preparation	0.10	0.10	630	630
Glaze preparation	0.10	0.10	827	827
Modeling and molds	0.10	0.10	4,133	4,133
Casting	0.10	0.10	9,551	3,184
Heating and ventilation	0.10	0.10	184	184
Drying	0.10	0.10	1,837	1,837
Glazing	0.10	0.10	4,592	4,592
Kilns	0.10	0.10	2,388	2,388
Sorting	0.10	0.10	1,837	1,837
Offices	0.10	0.10	918	918
Ware house	0.10	0.10	918	918
Facilities*	0.10	0.10	2,223	2,344
Total regulation lighting=			30,886	24,639
*Facilities split up				
Compressors	0.10	0.10	73	0
Boilers	0.10	0.10	18	18
Maintenance work shop	0.10	0.10	276	276
Losses hopper and crushing machine	0.10	0.10	157	157
Pumps and reverse osmosis plant	0.10	0.10	73	220
Wastewater plant	0.10	0.10	110	157
Security room	0.10	0.10	22	22
Waste treatment	0.10	0.10	4	4
Lunch room	0.10	0.10	37	37
Canteen	0.10	0.10	15	15
External lighting	0.10	0.10	1,181	1,181
Plugs circuit				
Watering				
Electrical station	0.10	0.10	110	110
Gas station	0.10	0.10	31	31
Generator	0.10	0.10	37	37
Water tank reservoir	0.10	0.10	79	79

Table 7.5 Summary of savings from generalized actions

	Annual savings	Plaster mold factory					Resin Mold Factory				
		kWh/year					kWh/year				
	Facilities	Efficiency	Frequency	LED	Regulation	Total	Efficiency	Frequency	LED	Regulation	Total
DB01	Raw Materials	9,317	36,749	1,790	850	48,706	9,317	36,749	1,790	850	48,706
DB02	Slip Preparation	22,661	282,924	1,326	630	307,541	22,661	282,924	1,326	630	307,541
DB03	Glaze Preparation	7,222	92,346	2,175	827	102,570	7,222	92,346	2,175	827	102,570
DB04	Modeling and Molds	15,405	0	10,875	4,133	30,413	677	0	10,875	4,133	15,685
DB05	Casting	71,606	391,664	25,133	9,551	497,954	27,591	78,464	8,378	3,184	117,617
DB06	Heating and Ventilation	3,197	37,723	483	184	41,587	3,197	37,723	483	184	41,587
DB07	Drying	28,823	149,384	4,833	1,837	184,876	28,823	149,384	4,833	1,837	184,876
DB08	Glazing	4,857	0	12,083	4,592	21,532	4,857	0	12,083	4,592	21,532
DB09	Kilns	14,910	243,900	6,283	2,388	267,481	14,910	243,900	6,283	2,388	267,481
DB10	Sorting	3,363	0	4,833	1,837	10,033	3,363	0	4,833	1,837	10,033
DB11	Offices	971	9,054	2,417	918	13,360	971	9,054	2,417	918	13,360
DB12	Ware house	0	0	2,417	918	3,335	0	0	2,417	918	3,335
DB13	Facilities	19,589	365,550	5,031	2,223	392,394	17,713	342,916	5,517	2,344	368,490
	Total Motor Efficiency	201,922	1,609,294	79,680	30,886	1,921,782	141,303	1,273,461	63,410	24,639	1,502,813

7.2.1 Increase of Thermal Insulation of the Mold Dryer

This measure is only applicable to the plaster mold factory as the equipment is not present when resin molds are used. From Table 6.3, thermal losses of the dryer are obtained and from them, loss reduction can be computed as follows:

$$\Delta Q_{dryerinsulation} = n_{dryers} \cdot Q_{losses} \cdot \frac{1}{f_{insulation}} \cdot \frac{n_{hours}}{day} \cdot \frac{n_{days}}{year} \quad (7.1)$$

$$3\text{dryers} \cdot 16.64 \frac{\text{kW}}{\text{year}} \cdot \frac{1}{2} \cdot 24 \frac{\text{hours}}{\text{day}} \cdot 232 \frac{\text{days}}{\text{year}} = 138,981 \frac{\text{kWh}}{\text{year}} \quad (7.2)$$

where:

$\Delta Q_{\text{dryer insulation}}$: thermal saving by insulation increase (kWh/year)

n_{dryers}: number of required dryers

Q_{losses}: thermal losses through enclosure for a dryer (kW).

7.2.2 Increase of Thermal Insulation of the Casting Hall

As described in the previous section, the walls and roof of the casting hall may be insulated with an extra thickness, which reduces the heat transfer coefficient. Table 5.18 states the thermal losses throughout walls and roof. During midday time, there is a strong influence of the solar radiation, which turns back the losses inward due to the increase in temperature on the roof. This fact affects to roof losses of the plaster factory during casting period and the averaged period of a resin mold factory. The greatest value of losses occurs during drying period in a plaster factory when their value corresponds to 58.93 kW through the walls and 98.16 through the roof. As an example, let us assume that insulation level is doubled, which would halve the losses. Hence, the energy saving would equal:

$$\frac{(58.93 + 98.16)\text{kW}}{2} \cdot \frac{14\text{h}}{\text{day}} \cdot \frac{232\text{days}}{\text{year}} = 255,100 \frac{\text{kWh}}{\text{year}}. \quad (7.3)$$

7.2.3 Thermal and Humidity Control of the HVU's

From Tables 5.18 and 5.19, the amount of air to be blown during the casting period at 25 °C is 93,072 kg just to evaporate water from ware. It takes 21 kW. On the contraire, during drying period, they are required 58,271 kg of air for water from ware and 677,303 for water from mold.

So, when no regulation is supplied, there is an excess of heat for extra ventilation during casting period corresponding with:

$$\frac{21.43\text{kW}}{93,072\text{kg}} \cdot [(58,271 + 677,303) - 93,072]\text{kg} \cdot \frac{10\text{h}}{\text{day}} \cdot \frac{232\text{days}}{\text{year}} = 343,221\frac{\text{kWh}}{\text{year}}. \tag{7.4}$$

7.2.4 *Increase of Thermal Insulation of the Ware Dryer*

From Table 5.21, losses throughout enclosures are 36.57 kW averaged for the whole cycle. Following the same procedure of 1.2.1 and 1.2.2, the calculation of losses reduction of the ware dryer is:

$$\frac{36.57\text{kW}}{2} \cdot \frac{24\text{h}}{\text{day}} \cdot \frac{232\text{days}}{\text{year}} = 101,809\frac{\text{kWh}}{\text{year}}. \tag{7.5}$$

7.2.5 *Thermal and Humidity Control of the Ware Dryer*

As explained in 1.2.3, the amount of air blown into the dryer might be controlled as per the requirement at any stage. Under such circumstances, the heat demand to warm up the air would be decreased. Table 7.6 states the required air flows for each drying stage versus actual ones in the case that no flow control is carried out.

By difference of averaged powers, it is calculated the excess of energy required for unnecessary blown air.

Table 7.6 Difference of inlet amount of drying air into the ware dryer

Savings by hygrometric control of the ware dryer						
Stage	Without flow control			With flow control		
	kg_{air}	kJ	kW	kg_{air}	kJ	kW
1	26,061	325,757	45	26,061	325,757	45
2	26,061	651,513	60	23,503	587,574	54
3	26,061	1,303,026	181	10,352	517,581	72
4	26,061	1,693,934	157	4,717	306,633	28
Total	104,242	3,974,230	110	64,633	1,737,545	48

$$(110.40 - 48.27)\text{kW} \cdot \frac{10\text{h}}{\text{day}} \cdot \frac{232\text{days}}{\text{year}} = 144,142\frac{\text{kWh}}{\text{year}}. \tag{7.6}$$

7.2.6 Load Lightening by SiC of the Tunnel Kiln

From (6.24), the energy to heat up cordierite is 3.48 times the required one for SiC. Thus, annual potential savings from material substitution would be calculated as:

$$Q_{\text{SiCtunnel}} = \frac{P_{\text{year}}}{\frac{n_{\text{pieces}}}{\text{cart}}} \cdot \frac{m_{\text{cordierite}}}{\text{cart}} \cdot C_{\text{cordierite}} \cdot \Delta T \cdot \frac{1\text{ h}}{3,600\text{ s}} \cdot \left(1 - \frac{1}{3.48}\right) \tag{7.7}$$

where:

$Q_{\text{SiC tunnel}}$: annual energy reduction (kWh/year)

P_{year}: annual production (pieces/year) from Table 3.1

n_{pieces}: number of pieces per cart from Fig. 4.24

$m_{\text{cordierite}}$: mass of cordierite on the cart (kg) from Fig. 4.27

$C_{\text{cordierite}}$: specific heat of the cordierite (kJ/kg/K) from Table 6.11

ΔT: difference of temperatures of the cordierite (K) from Fig. 4.21.

$$\frac{500,000\frac{\text{pieces}}{\text{year}}}{18\frac{\text{pieces}}{\text{cart}}} \cdot 255\frac{\text{kg}_{\text{cordierite}}}{\text{cart}} \cdot 1.46\frac{\text{kJ}}{\text{kg}\cdot\text{K}} \cdot (1,250 - 25) \cdot \frac{1\text{ h}}{3,600\text{ s}} \cdot \left(1 - \frac{1}{3.48}\right)$$
$$= 2,508,224\frac{\text{kWh}}{a\,no}. \tag{7.8}$$

7.2.7 Load Lightening by SiC of the Shuttle Kiln

As calculated for the tunnel kiln, the savings for the shuttle kiln would be computed as:

$$Q_{\text{SiCshuttle}} = \frac{m_{\text{cordierite}}}{\text{cycle}} \cdot \frac{\text{cycles}}{\text{year}} \cdot C_{\text{cordierite}} \cdot \Delta T \cdot \frac{1\text{ h}}{3,600\text{ s}} \cdot \left(1 - \frac{1}{3.48}\right) \tag{7.9}$$

where:

$Q_{\text{SiC shuttle}}$: annual energy reduction (kWh/year)

$m_{\text{cordierite}}$: mass of cordierite in the kiln each cycle (kg) from Table 5.28

$C_{\text{cordierite}}$: specific heat of the cordierite (kJ/kg/K) from Table 6.11

ΔT: difference of temperatures of the cordierite (K) from Fig. 4.36

$$3,994\frac{\text{kg}_{\text{cordierite}}}{\text{cycle}} \cdot 300\frac{\text{cycle}}{\text{year}}1.46\frac{\text{kJ}}{\text{kg}\cdot\text{K}} \cdot (1,206-25) \cdot \frac{1\ \text{h}}{3,600\ \text{s}} \cdot \left(1-\frac{1}{3.48}\right) = 409,046\frac{\text{kWh}}{a\ no}. \quad (7.10)$$

7.2.8 Heating Recovery from Compressor Refrigeration

As stated in Sect. 6.10.2, up to 128 thermal kW can be obtained from the refrigeration of the compressor. The corresponding annual savings for the set of two compressors based on a duty cycle of 2,225 h are calculated as follows:

$$2\text{compressors} \cdot 128\text{kW} \cdot 2,225\frac{\text{h}}{\text{year}} = 569,600\frac{\text{kWh}}{\text{year}}. \quad (7.11)$$

7.2.9 Summary of Thermal Non-Exclusive Savings

In the same line that electrical savings, thermal ones may be summarized in Table 7.7.

Possibilities of saving are greater in a plaster mold factory because of molds making and hall heating thermal demands.

Table 7.7 Summary of thermal savings. (*Source* Data compiled by the authors)

Thermal savings	PMF	RMF
	kWh/year	kWh/year
Increase of thermal insulation of the mold dryer	138,981	
Increase of thermal insulation of the casting hall	255,100	87,786
Thermal and humidity control of the HVU's	343,221	
Increase of thermal insulation of the ware dryer	101,809	101,809
Air flow variation by temperature and humidity control	144,142	144,142
Load lightening by SiC of the tunnel kiln	2,508,224	2,508,224
Load lightening by SiC of the shuttle kiln	409,046	409,046
Heating recovery from compressor refrigeration	569,600	569,600
Total saving=	4,470,122	3,820,606

7.3 Exclusive Thermal Savings

7.3.1 Heating Recovery

From Table 5.15, it has been built that Table 7.8 summarizes the demand of thermal consumers and the corresponding level of temperature.

On the other hand, from Table 5.26, the annual recoverable heat from the tunnel kiln equals:

$$Q_{\text{recovery}} = 1,681\text{KW} \cdot 24\frac{\text{hours}}{\text{day}} \cdot 300\frac{\text{days}}{\text{year}} = 12,106,582\frac{\text{kWh}}{\text{year}}. \quad (7.12)$$

In principle, it seems that the kiln is able to supply all the rest of thermal consumers. However, their demand is not constant. Hence, there will be periods with excess of available heat and other ones with lack of it as will be analyzed in Chap. 8. Besides, the tunnel kiln has been set at a duty schedule of 300 days per year, while other consumers may be operated at a rate of 232 days per year.

HVU and ware dryers are quite variable regarding power consumption. The greater demand occurs when HVU is operating in drying mode at night in a plaster mold factory and the ware dryer is at the third stage. Table 7.9 shows comparatively how the thermal demand varies depending on the time and the type of factory and how it affects to the capability of the kiln to supply completely to the consumers.

During the drying period of the PMF, there is a lack of power that equals:

$$Q_{\text{lack}} = (2,276 - 1,681)\ \text{kW} = 595\ \text{kW}. \quad (7.13)$$

On the other hand, there is an excess in other periods or factory type. It might be worth considering a heat storage system.

Table 7.8 Thermal consumers which may be supplied from recovered heat

Heat recovery potential				
Facilities	PMF		RMF	
	W_{year}	T	W_{year}	T
	kWh/year	°C	kWh/year	°C
Molds dryers	2,004,792	60		
Boilers			1,118,246	45
Heating and ventilation unit	4,351,851	25–40	609,956	25
Ware dryers	2,112,524	90	2,112,525	90
Boilers	581,856	45	581,856	45
Total requirement=	9,051,023		4,422,583	

Table 7.9 Heat recovery balance under the greater thermal demand

Heat recovery potential			
	PMF	PMF	RMF
Facilities	Casting	Drying	Casting
	kW	kW	kW
Molds dryers	153	153	153
Boilers at casting hall			603
Heating and ventilation unit	23	1,323	110
Ware dryers	695	695	695
Boilers at facilities	105	105	105
Total Requirement=	976	2,276	1,665

7.3.2 Induced Draught for Tunnel Kiln

As stated in Sect. 6.9.1, it can be created an induced draught to bring the hot air upward inside the kiln chamber in order to provide heat to the load instead of being delivered to the chimney. Table 7.8 shows how the recoverable heat is lower so as the temperature with regard to the conventional kiln (Table 7.10).

On the other hand, the thermal inlet requirement is lower as stated in Table 7.11 because of the decrease in flow gases through the kiln.

The fuel consumption results in 21% less than conventional kiln.

Table 7.10 Distribution of thermal power through the stacks when applied induced draught. (*Source* Data compiled by the authors)

Exhausting per stack				
Stack	Fluid	m	T	Q
		kg/s	K	kW
Exhausting	$H_{CO2\ OUT}$	0.0910	355	6
	$H_{H2O\ OUT}$	0.0745	355	198
	$H_{air\ excess\ OUT}$	0.0568	355	5
	$H_{N2\ OUT}$	0.4333	355	37
	$H_{roof\ air\ OUT}$	0.3120	355	26
	$H_{draught\ OUT}$	0.0749	355	6
	$H_{rapid\ cooling\ OUT}$	0.6553	355	54
Slow cooling	$H_{slow\ cooling\ OUT}$	2.9677	576	916
Recoverable heat		4.6655	496	1,248

Table 7.11 Thermal requirement of the kiln with induced draught

Thermal requirements		
Q	kW	%out
Q_{ware}	34	2.15
Q_{load}	31	1.97
Q_{cart}	84	5.31
$Q_{exhausting}$	288	18.12
Q_{sc}	842	53.00
Q_{walls}	135	8.48
Q_{root}	144	9.05
Q_{floor}	30	1.92
Total	1,589	100

7.3.3 Primary Air Preheating for Tunnel Kiln

In Table 7.12, it can be seen the new distribution of recovered heat for the new configuration.

371 kW from slow cooling air outlet is devoted to pre-heat combustion and roof air. Hence, the final recoverable heat gets reduced by 58% with regard to conventional kiln.

Table 7.13 shows the new thermal requirement for the optimized configuration.

Paradoxically, exhausting requirement results negative. The explanation lies in the fact that inlet air is hotter than final exhausting through the stack. Consequently, thermal requirement for the kiln would be cut by 46%.

Table 7.14 summarizes the improvements achieved by inducing draught and preheating inlet airs.

Table 7.12 Distribution of thermal power through the stacks of the optimized kiln

Exhausting per stack				
Stack	Fluid	m	T	Q
		kg/s	K	kW
Exhausting	HCO_2 OUT	0.0624	316	2
	HH_2O OUT	0.0511	316	132
	Hair excess OUT	0.0390	316	2
	HN_2 OUT	0.2988	316	13
	H_{roof} air OUT	0.3120	316	13
	$H_{draught}$ OUT	0.0749	316	3
	H_{rapid} cooling OUT	0.6553	316	28
Slow cooling	H_{slow} cooling OUT recoverable	2.2956	492	509
Recoverable heat		3.7891	423	703
	H_{slow} cooling OUT pre-heating	0.6721	803	371

Table 7.13 Distribution of the thermal requirements of the optimized kiln

Thermal Requirements		
	kW	%
Q_{ware}	34	3.14
Q_{load}	31	2.88
Q_{cart}	84	7.74
$Q_{exhausting}$	−176	−16.11
Q_{sc}	806	73.99
Q_{walls}	135	12.37
Q_{roof}	144	13.19
Q_{floor}	30	2.80
TOTAL	1,089	100.00

Table 7.14 Comparative performance of the different tunnel kiln configurations

Kiln configuration	Q_{NG}	W_{NG}	$m_{recoverable}$	$T_{recoverable}$	$Q_{recoverable}$	$W_{recoverable}$
	kW	kWh/year	kg/s	°C	kW	kWh/year
Conventional kiln	2,017	14,524,898	4.8455	288	1,681	12,106,582
Kin with induced draught	1,589	11,438,599	4.6655	223	1,248	8,986,072
Kiln optimized with induced draught and pre-heated air	1,089	7,844,158	3.7891	150	703	5,064,305

7.3.4 Primary Air Preheating for Shuttle Kiln

Following the model to study the shuttle kiln in Sect. 4.9.2 and the results for the conventional configuration in Sect. 5.2.5, it has been built Table 7.15 to compare the savings that can be achieved by pre-heating the inlet air during heating cycle. A substantial reduction of fuel is obtained; though residual heat is decreased for this purpose.

Efficiency of the kiln improves by 41%.

7.3.5 Condensation by Absorption Machine

In order to allow for the calculation, thermodynamic model has been considered for the system as it was done in Chap. 4 for common equipment. In this sense, the computational modeling allows the researcher to make predictions about the results that could be tested later on in a second stage of research [1]. In the first stage, the results have to be coherent with existing data. In the area of hybrid simulation [2], the research provides a tool to better evaluate the relationship between actual existing

Table 7.15 Comparative performance of a conventional and optimized shuttle kiln

Shuttle kiln optimization			
Heating cycle		Conventional	Pre-heating
Thermal fuel requirement	kJ	51,310,449	28,132,341
Duty cycle	min	910	910
Averaged power for the cycle	kW	940	515
Pre-heating requirement	kJ		11,966,654
Residual heat	kg/cycle	85,763	14,733
Residual heat temperature	K	755	731
Residual heat energy	kJ/cycle	49,487,712	10,888,928
Annual energy recoverable	kWh/year	1,626,931	907,411
O_2 content	%	16.6694	18.1356
Humidity content	kg_{water}/kg_{air}	0.02554	0.01689

data in the sanitary-ware manufacturing and the parameters that allow to optimize the performance of the efficiency of resources.

Environmental drying performance of the air depends on the psychrometric characteristics of the ambient air whose annual averaged values were defined in Sect. 5.2.2.

As stated in Sect. 6.12.2, the approach of this case study is to recover evaporated water contained in the exhausting of the different elements of the equipment. The system is based on an absorption machine producing cool to condensate evaporated water within exhausting. There are some examples of research regarding industrial applications of absorption machines operated with residual heat [3]. The proposed configuration is shown in Fig. 7.1

Recovered heat from kiln with high content in oxygen and humidity at a relative high temperature is mixed with fresh air at the hygrometric requirement for MD, WD or HVU. This mixture is blown into the chamber of the dryers or casting hall and absorbs water from ware or mold, which is condensed before released to the external atmosphere through the absorption cooling. Besides, the refrigeration of the absorption cycle is provided by fresh air at once it is pre-heated before mixing with recovered fumes from kiln. The absorption machine model selected for this case study is based in a one-stage conventional commercial lithium bromide machine corresponding with Fig. 7.2.

By means of mass and energy balances applied to the consumers according to their thermodynamic model and considering the following system of equations, it is computed whether the recovered heat from the kiln is able to supply all the consumptions demanded.

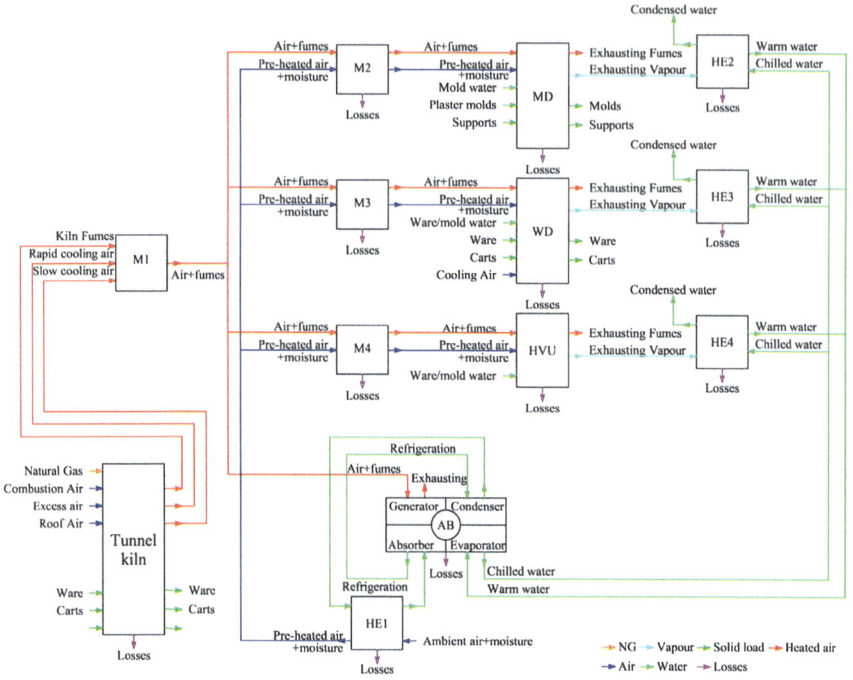

Fig. 7.1 Configuration of the water recovery by absorption cycle

7.3.5.1 Psychrometric Model

The saturation pressure of the vapor at a given temperature is calculated by Sonntag's formula [4]:

$$p_s = \exp\left(\sum_{i=1}^{4} a_i \cdot T^{i-2} + a_7 \ln(T)\right), \tag{7.14}$$

where:

$a1=$	−6.09693850E+03
$a2=$	2.12409642E+01
$a3=$	−2.71119300E-02
$a4=$	1.67395200E-05
$a7=$	2.43350200E+00
p_s	(Pa)
T	(K)

with a standard uncertainty < 0.005%rh.

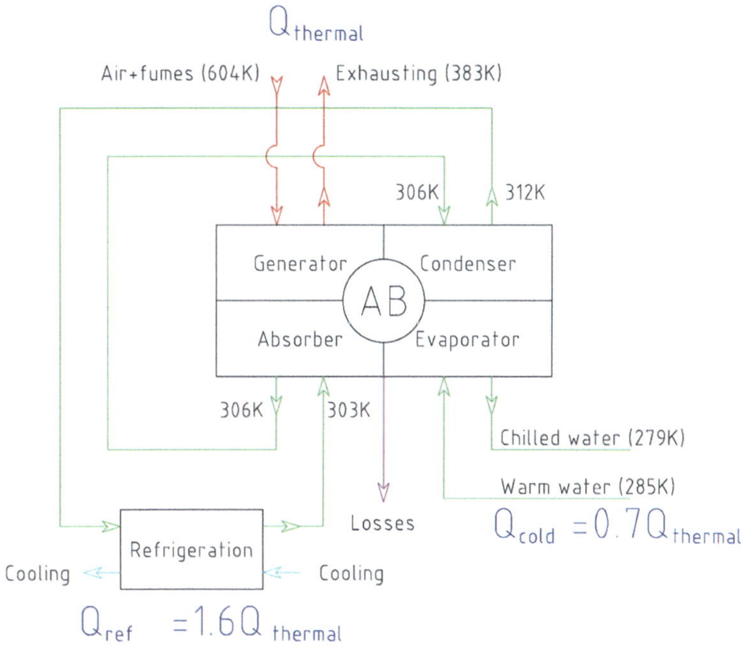

Fig. 7.2 Absorption model scheme

Humidity ratio of moist air can be expressed as:

$$x = \frac{m_v}{m_a} \cdot \frac{\text{kg}_{\text{vapor}}}{\text{kg}_{\text{air}}} \tag{7.15}$$

where:

x = Water contain (kg/kg).

m_v = Vapor mass (kg).

m_a = Air mass (kg).

Correspondent masses of vapor and air are expressed as:

$$m_v = \frac{M_w \cdot p_v \cdot V}{R_0 \cdot T}; \tag{7.16}$$

$$m_a = \frac{M_a \cdot p_a \cdot V}{R_0 \cdot T} \tag{7.17}$$

where:

M_w = molecular mass of vapor (kg/kmol).

p_v = partial pressure of vapor (Pa).

V = volume of the vapor-air mixture (m^3).

R_0 = ideal gas constant (kJ/K/kmol).

T = temperature of the mixture (K).

M_a = molecular mass of air (kg/kmol).

p_a = partial pressure of air (Pa).

Hence:

$$x = \frac{m_v}{m_a} = \frac{M_v \cdot p_v}{M_a \cdot p_a}. \tag{7.18}$$

Relative humidity can be expressed as:

$$\phi = \frac{p_v}{p_s} \tag{7.19}$$

and total pressure of the mix:

$$p = p_v + p_a. \tag{7.20}$$

Thus, combining (5), (6) and (7):

$$x = \frac{M_v}{M_a} \cdot \frac{p_s}{\frac{p}{\phi} - p_s} \frac{\mathrm{kg_{vapor}}}{\mathrm{kg_{air}}}. \tag{7.21}$$

At saturation pressure,

$$\phi = 1 \tag{7.22}$$

and

$$x_s = \frac{M_v}{M_a} \cdot \frac{p_s}{p - p_s} \frac{\mathrm{kg_{vapor}}}{\mathrm{kg_{air}}}. \tag{7.23}$$

Then, actual water content is provided by relative humidity of the environment conditions with:

$$x = x_s \cdot \phi \cdot \frac{\mathrm{kg_{vapor}}}{\mathrm{kg_{air}}}. \tag{7.24}$$

Evaporation energy is provided by thermal energy, either by combustion of NG or dry heated air. Hence, the mathematical problem is solved by combining mass and energy balances applied to every system:

$$\sum_{\text{in}} \dot{m}_{\text{in}} = \sum_{\text{out}} \dot{m}_{\text{out}} \tag{7.25}$$

$$\sum_{\text{in}} \dot{m}_{\text{in}} \cdot h_{\text{in}} = \sum_{\text{out}} \dot{m}_{\text{out}} \cdot h_{\text{out}}. \tag{7.26}$$

Enthalpies of flows are obtained by the NASA Glenn Research Center database of thermodynamic data JANAF coefficients

7.3.5.2 Dryers and HVU Calculation

The aim of both processes is essentially the same, to evaporate the water contained in the ware and mold by blowing fresh air relatively dry and warm into the chamber or the hall. The way of calculation and the system of equations is basically the same for each of the systems with the only difference of set points and the source of fresh air. For a HVU, fresh air is taken from outside the factory. For the dryers, fresh air is taken from the hall where the equipment is located.

In order to solve the balances, it is considered that the thermal requirement has to:

Heat up the Ware and the Supporting Carts

$$Q_{\text{carts\&ware}} = (m_{\text{ware}} \cdot Ce_{\text{ware}} + m_{\text{carts}} \cdot Ce_{\text{carts}}) \cdot \Delta t \tag{7.27}$$

where:

m_{ware} = mass of ware in the chamber

Ce_{ware} = specific heat for the ware

m_{ware} = mass of carts in the chamber

Ce_{ware} = specific heat for the carts

Δt = increase of temperature as per each stage of drying process

Heat the Fresh Air to Absorb Evaporated Water

$$Q_{\text{ventilation}} = m_{\text{freshair}} \cdot Ce_{\text{air}} \cdot \Delta t \tag{7.28}$$

where:

m_{freshair} = mass of fresh air needed to evaporate

Ce_{air} = specific heat for the air at required temperature

Δt = temperature difference between inside and outside of the chamber.

The needed amount of fresh air taken from either the hall or external is calculated by:

$$m_{\text{freshair}} = \frac{m_{\text{evaporation}}}{xs_{\text{IN}} - x_{\text{IN}}} \tag{7.29}$$

where:

$m_{\text{evaporation}}$ = mass of water to be evaporated in each stage of the process. It can be taken from Table 4.16.

xs_{IN} = moisture saturation of air at the chamber condition

x_{IN} = moisture of air at the chamber condition

Keep the Temperature into the Chamber Because of the Thermal Losses

$$Q_{\text{losses}} = S \cdot U \cdot \Delta t \tag{7.30}$$

where:

S = surface of chamber enclosure

U = global coefficient of heat transmission

Δt = temperature difference between inside and outside.

Evaporate the Water to Be Extracted

$$Q_{\text{evaporation}} = m_{\text{evaporation}} \cdot \left(h_{\text{vapor}} - h_{\text{water}}\right) \tag{7.31}$$

where:

h_{vapor} = specific enthalpy of vapor at dryer condition

h_{water} = specific enthalpy of water at dryer condition.

The total thermal requirement for the system is:

$$Q_{\text{thermal}} = Q_{\text{carts\&ware}} + Q_{\text{ventilation}} + Q_{\text{losses}} + Q_{\text{evaporation}}. \tag{7.33}$$

In order to compute the final water to be recovered from exhausting, they have to be calculated:

Mass of Water from Fresh Air

When fresh air is supplied to the system, the moisture contained in it is calculated by:

$$m_{\text{moisturefromfreshair}} = m_{\text{freshair}} \cdot x_{\text{OUT}} \tag{7.33}$$

where:

x_{OUT} = moisture of supplied fresh air.

Mass of Water from Combustion Air When Air Is Fueled by Natural Gas

The amount of water contained in combustion fresh air is calculated as:

$$m_{\text{moisturecombustionair}} = m_{\text{combustionair}} \cdot x_{\text{OUT}}. \tag{7.34}$$

Mass of Water from Recovered Heat

When dryer is supplied by recovered heat, there is a mixer (M1) where the hot flow is cool down up to the dryer level requirement with fresh air from the hall. The ratio of recovered heat and fresh air, so as the resultant water content is calculated by the following system of equations:

From an energy balance:

$$m_{RH} \cdot h_{RH} + m_{\text{OUT}} \cdot h_{\text{OUT}} = Q_{\text{thermal}} + (m_{RH} + m_{\text{OUT}}) \cdot h_{\text{IN}}. \tag{7.35}$$

From a mass balance of water content:

$$m_{RH} \cdot x_{RH} + m_{\text{OUT}} \cdot x_{\text{OUT}} = (m_{RH} + m_{\text{OUT}}) \cdot x_{\text{IN}}. \tag{7.36}$$

From the mix of recovered heat (RH) required to absorb the evaporated water of the ware:

$$m_{\text{evaporated}} = (m_{RH} + m_{\text{OUT}}) \cdot (xs_{\text{IN}} - x_{\text{IN}}). \tag{7.37}$$

The moisture contained in it is calculated by:

$$m_{\text{moistureflow}} = m_{RH} \cdot x_{RH} + m_{\text{OUT}} \cdot x_{\text{OUT}}. \tag{7.38}$$

System Results

The system of equations resolved for each thermal consumer yields the mass and energy balance for every one of them as shown in the following tables. These balances are calculated under the basis of recovered heat as fuel supply. The balances provided by natural gas were calculated in Sect. 5.2 (Tables 7.16, 7.17, 7.18 and 7.19).

Table 7.16 Balance for the mold dryer when fueled with recovered heat. (*Source* Data compiled by the authors)

Facility	Flow	m_{IN}	H_{IN}	m_{OUT}	H_{OUT}		Q_{fuel}	Q_{demand}	
		kg	kJ	kg	kJ		kJ	kJ	%
MD	RH CO_2	10,189	2,777,941	10,189	518,849	Q_{RH}	237,262,003		
	RH H_2O	8,337	25,446,101	8,337	21,782,588				
	RH N_2	48,785	14,722,225	48,785	3,043,594				
	RH air	943,413	276,552,702	943,413	56,891,935				
	Psychrometric air	637,109	15,994,481	637,109	38,420,475	Q_{air}		22,425,994	9.45
	Water of molds	72,829	7,634,651	72,829	190,290,897	$Q_{vaporization}$		182,656,246	76.99
	Supports	24,581	320,787	24,581	769,889	$Q_{supports}$		449,102	0.19
	Molds	54,941	1,098,821	54,941	2,637,170	Q_{molds}		1,538,349	0.65
	Losses	0	0	0	30,192,342	Q_{losses}		30,192,342	12.73
	Total=	1,800,184	344,547,709	1,800,184	344,547,739		237,262,003	237,262,033	100

Table 7.17 Balance for the ware dryer when fueled with recovered heat. (*Source* Data compiled by the authors)

Facility	Flow	m_{IN}	H_{IN}	m_{OUT}	H_{OUT}		Q_{fuel}	Q_{demand}	
		kg	kJ	kg	kJ		kJ	kJ	%
WD	RH CO_2	1,395	380,421	1,395	68,938	Q_{RH}	12,522,705		
	RH H_2O	1,142	3,484,681	1,142	2,979,296				
	RH N_2	6,681	2,016,115	6,681	404,812				
	RH air	49,282	14,446,615	49,282	2,886,357				
	Ambient air	42,910	1,077,254	42,910	2,513,169				
	Ambient moist	479	1,220,968	479	1,250,776				
	Water of ware	4,286	449,333	0	0	Q_{water}		−449,333	−3.59
	Ware	21,550	450,395	21,550	450,395	Q_{ware}		0	0.00
	Carts	12,800	167,040	12,800	167,040	Q_{carts}		0	0.00
	Cooling air	48,183	1,209,619	48,183	2,821,971	$Q_{cooling\ air}$		1,612,351	12.88
	Vaporized water	0	0	4,286	11,185,597	$Q_{evaporation}$		11,185,597	89.32
	Losses	0	0	0	174,078	Q_{losses}		174,078	1.39
	Total=	188,709	24,902,442	188,709	24,902,431		12,522,705	12,522,694	100

Table 7.18 Balance for the HVU for the PMF when fueled with recovered heat. (*Source* Data compiled by the authors)

Facility	Flow	m_{IN}	H_{IN}	m_{OUT}	H_{OUT}		Q_{fuel}	Q_{demand}	
		kg	kJ	kg	kJ		kJ	kJ	%
HVU PMF	RH CO_2	4,878	1,552,764	4,878	161,953	Q_{RH}	43,401,198		
	RH H_2O	3,991	12,525,392	3,991	10,275,991				
	RH N_2	23,356	8,120,995	23,356	961,687				
	RH air	172,287	58,239,545	172,287	6,854,725				
	Ambient air mix	754,721	11,493,709	754,721	30,027,829				
	Ambient moist mix	5,464	13,818,099	5,464	14,067,121				
	Ware vaporizing water	1,724	286,168	1,724	4,439,118	Q_{ware}		4,152,950	9.57
	Mold vaporizing water	11,690	1,940,314	11,690	30,098,669	Q_{mold}		28,158,355	64.88
	Losses	0	0	0	11,089,905	Q_{losses}		11,089,905	25.55
	Total=	978,111	107,976,987	978,111	107,976,999		43,401,198	43,401,210	100

Table 7.19 Balance for the HVU for the RMF when fueled with recovered heat. (*Source* Data compiled by the authors)

Facility	Flow	m_{IN}	H_{IN}	m_{OUT}	H_{OUT}		Q_{fuel}	Q_{demand}	
		kg	kJ	kg	kJ		kJ	kJ	%
HVU RMF	RH CO_2	582	185,278	582	12,086	Q_{RH}	7,095,306		
	RH H_2O	476	1,494,543	476	1,213,160				
	RH N_2	2,787	969,006	2,787	72,422				
	RH air	20,557	6,949,206	20,557	516,091				
	Ambient air mix	68,828	1,048,188	68,828	1,727,912				
	Ambient moist mix	505	1,276,399	505	1,285,644				
	Ware vaporizing water	431	45,185	431	1,098,029	Q_{ware}		1,052,844	14.84
	Mold vaporizing water	0	0	0	0	Q_{mold}		0	0.00
	Losses	0	0	0	6,042,464	Q_{losses}		6,042,464	85.16
	Total=	94,166	11,967,805	94,166	11,967,807		7,095,306	7,095,307	100

Table 7.20 Thermal power and mass flow exhausting in a plaster mold factory for dryers and HVU

Plaster mold factory					
IN	m_{IN}	H_{IN}	OUT	m_{OUT}	H_{OUT}
	kg/s	kW		kg/s	kW
RH	5.5155	1,973	Flue gases + Air	17.6044	1,053
Air	12.0889	217	Vapor	0.4593	1,191
Moisture	0.1388	351			
Water	0.3839	50			

Table 7.21 Thermal power and mass flow exhausting in a resin mold factory for dryers and HVU

Resin mold factory					
IN	m_{IN}	H_{IN}	OUT	m_{OUT}	H_{OUT}
	kg/s	kW		kg/s	kW
RH	1.7597	624	Flue gases + Air	4.8567	336
Air	3.0970	70	Vapor	0.1312	342
Moisture	0.0066	17			
Water	0.1132	12			

These balances can be compared with Table 5.25, where the available recovered heat was summarized. Available heat to be recovered from kiln is 4.8455 kg/s carrying 1,681 kW.

In order to find out if there is recovered heat enough for every consumer, it is needed to scale every consumption to the same period to be able to compare them each other. Mold dryers operate in a cycle of 7 days, ware dryers in 11 h and HVU's in 24 h. Since the kiln operates at the same rate on continuous basis, it will be calculated the power and the mass flow per second to make the comparison. Tables 7.20 and 7.21 show resultant the values of inlet and outlet flows of dryers and HVU for the two types of factory. These values define the amount of recovered heat required for evaporation and the final exhausting vapor that can be condensed by the absorption cycle.

In the case of the plaster mold factory, there is a lack of

$$Q = (5.5155 - 4.8455)\frac{\text{kg}}{\text{s}} = 0.67\frac{\text{kg}}{\text{s}}. \tag{7.39}$$

That has to be supplied with natural gas.

On the other hand, in the resin mold factory, there is an excess of recovered heat that is available for other heating purposes.

$$Q = (4.8455 - 1.7597)\frac{\text{kg}}{\text{s}} = 3.0858\frac{\text{kg}}{\text{s}}. \tag{7.40}$$

Table 7.22 Potential condensates from MD, WD and HVU

Water content in exhausting			
Plaster mold factory	*kg*	*Cycle*	*kg/day*
Mold dryer	81,166	7 days	11,595
Ware dryer	5,907	11 h	5,907
HVU PMF	22,869	24 h	22,869
		TOTAL=	40,371
Resin mold factory	*kg*	*Cycle*	*kg/day*
Ware dryer	5,907	11 h	5,907
HVU RMF	1,412	24 h	1,412
		TOTAL=	7,319

Other important finding from these balances is the amount of water that can be condensed from exhausting. Since the cycles are different,

- 7 days for the mold dryer
- 11 h/day for the ware dryer
- 24 h/day for the HVU.

They will be computed for a daily cycle. Table 7.22 summarizes the results.

From Table 7.21, it can be calculated the thermal power required for the absorption machine in both cases as per its coefficient of performance which is assumed as 0.7 according to Fig. 7.2.

$$Q_{\text{absorption}} = \frac{Q_{\text{condensation}}}{0.7} = \frac{m_{\text{vapor}} \cdot \left(h_{\text{vapor}} - h_{\text{liquide}}\right)}{0.7} \tag{7.41}$$

where:

$Q_{\text{absorption}}$: thermal power to fuel the absorption machine (kW)

$Q_{\text{condensation}}$: cooling power to condensate the vapor (kW)

m_{vapor}: mass flow of vapor to be condensed (kg/s)

h_{vapor}: specific enthalpy of the vapor (kJ/kg)

h_{liquid}: specific enthalpy of the water (kJ/kg).

7.3.6 MED Distillation

MED distillation optimization has been researched [5] and it has been demonstrated that by increasing the number of effects or temperature difference in the cells the efficiency of the MED is improved. Thus, the overall performance would be better. The total surface of heat transfer appears quite independent of the number of effects.

On the other hand, there are several main constraints that limit the number of effects, so as the design of the MED [6]:

- the higher the temperature of the heat source, the higher the GOR.
- exceeding the solubility limits of the brine inside the tubes would produce scaling and therefore, an important drop in performance. The top brine temperature is 70 °C in order to avoid anti-scalants.
- the brine temperature inside the last effect has to be at least 3 K greater than the feed temperature.
- the difference of brine temperature between two effects has to be greater than 2.25 K and smaller than 5 K.
- the temperature increase of the seawater inside the condenser has to be greater than 3 K and smaller than 10 K.
- and, of course the capital cost of the investment is an important limitation in the design in order to increase the number of effects.

MED distillation is the third water desalination technology with a share of 7% worldwide. Mostly, the treated water is destined for human consumption, thus the level quality is largely proved. Current distillation plants are able to supply distillate at purities of 5 to 1 ppm [7]. There is some research regarding the quality of the water, which meets the standards stated by WHO for drinking water. This is the level that distillation plants can achieve [8]

Constraints of feeding water regard the solubility limits of dissolved salts, especially $CaSO_4$, which are able to produce scaling. This is a factor to be taken into account when designing the system [9]. Nevertheless, a good maintenance routine may palliate it by regular cleansing with HCl.

The salinity that the system is able to handle rises up to 45,000 ppm of TDS

The model has been tested for a different number of effects to find the optimal one. It can be seen that only one more effect causes the draught of the brine and the consequent inefficiency of the system. Mathematically, efficiency seems to be 1 but it is theoretically calculated assuming. No water remains in the residual brine, which is not feasible. Liquid stream is needed to carry the dissolved salts. For the purpose of this case study, it has been taken a MED distiller with 11 effects.

In Sect. 6.12, Fig. 6.18 and Table 6.15 stated the amount of water to be treated at the water treatment plant.

As stated in [10], the kiln exhausting is able to supply recovered heat for the thermal requirement beyond the kilns in a ceramic sanitary-ware factory. When recovered heat is devoted to MED, the rest of thermal consumers cannot be supplied from kiln. Table 5.26 in Sect. 5.2.4.3 states the available recovered heat to supply the MED distiller. These values set the starting point to design the MED model.

A suitable strategy might be to combine a MED system to distillate water with a co-generation plant to provide energy. So, the recovered heat from the power plant might supply heat for the MED while the kiln is serving the rest of thermal consumers. This approach will be discussed in Chap. 9.

Distillation is less sensible to suspended solids content of water as far as the rejected brine keeps the particles suspended by a proper turbulent flow. On the other hand, it is suitable to improve heat transfer along the tubes of the heat exchangers. The water recycled with this method is suitable to be used in new cycles. The potential benefit of this method, besides the water saving, is the ability of taking energy from recovered heat from the kiln. Some research has been developed in this regard in the industrial sector [11]. Though great electrical power is needed to pump out the brine from the distiller, the advantage of re-using water yields a better global efficiency.

MED mathematical model is developed to find the right balance of energy from recovered heat through the stacks of the kiln and the thermal requirement of a distiller to remove salts and impurities from flushing water.

The core of this section is to find the way to take the most energy from the recovered heat of the kiln. Thus, the design of the distiller has to match the heat characteristics in order to run under low grade heat range [12]. It is important to keep the temperature of the water to feed the MED below 70 °C. Since gas exhausting temperature in the MED is to be over 80 °C, the available heating power is 1,097 kW due to the fact that the temperature difference is lower. Nevertheless, it has to be taken into account that 70 °C is the higher temperature at the evaporator tubes in order to avoid scaling at the brine side [13]. Over this temperature, some content of $CaCO_3$ or $CaSO_4$, might decant drastically producing scale and great drops of the heat transfer coefficient. Due to the low grade of the heat, spray flash chamber is advisable for the distiller [14]. On the other hand, the water to be treated is.

7.3.6.1 Cascade of Two Simple Evaporating Cells

Initially, as a pre-sizing exercise, it was tried a system with two boilers at the two different thermal states of the recovered gases from the stacks of the kiln. They were calculated at atmospheric pressure and making a thermal cascade deriving exhausting from the hotter towards lower temperature as Fig. 7.3 shows.

The equations for this model are obtained from energy balances in the evaporator and condenser of each stage as per Fig. 9:

$$Q_e = \dot{m}_g \cdot \Delta h_g = \dot{m}_v \cdot \Delta h_v \tag{7.42}$$

$$Q_c = \dot{m}_w \cdot \Delta h_w = \dot{m}_c \cdot \Delta h_c. \tag{7.43}$$

When trying a simple evaporator with three levels of temperature matching the three flows of gas from the stacks of the kiln, the efficiency results as bad as 13.87% for the PMF and 4.05% for RMF. This simulation suggests providing a more efficient configuration, which results in a MED distiller modified to meet the thermodynamic state of the source of heat improving notably the efficiency.

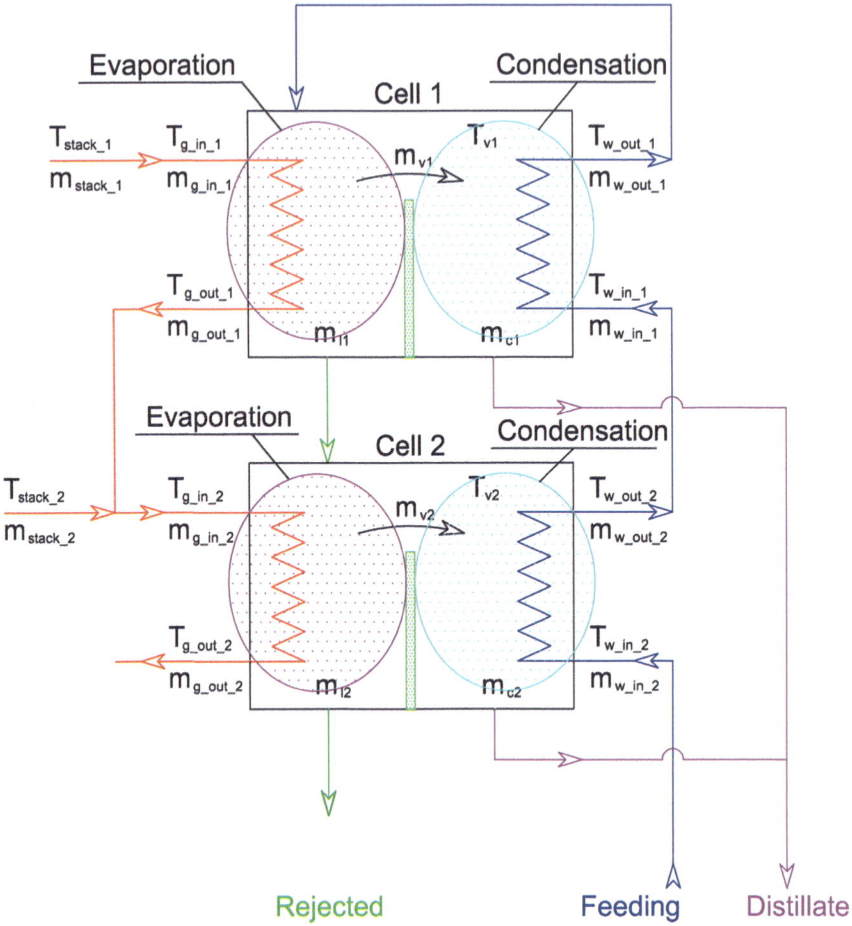

Fig. 7.3 Simple cascade of atmospheric evaporator-condenser cells

7.3.6.2 MED Model

In order to improve the performance, a vacuum boiling system in cascade was modeled as Fig. 7.4 shows. It has been conceived as a MED system.

Water feeding to be recycled is sent through the pre-heaters up to the first effect where it is sprayed onto its evaporator. Recovered heat from the kiln stacks provides energy to evaporate a fraction of this flow in which downward is condensed by feeding water coming from the second effect. The temperature of the inlet water is assumed to equal the temperature within the effect. Under such circumstances, no heat is devoted to heat up the liquid up to the boiling point. Schematic configuration of the first effect can be seen in Fig. 7.5.

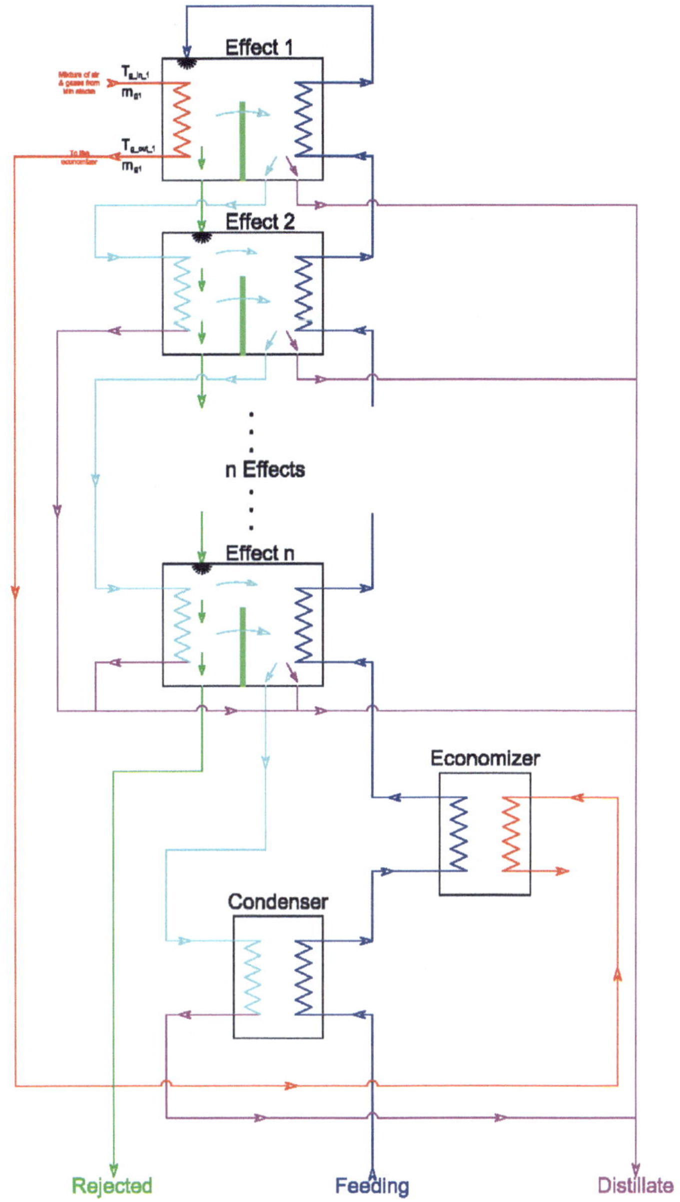

Fig. 7.4 MED model scheme

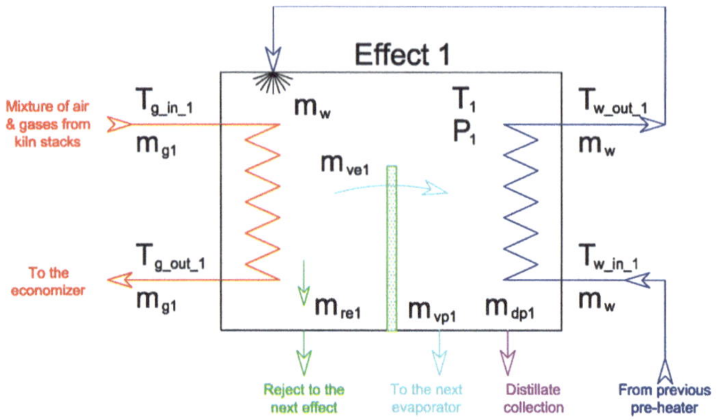

Fig. 7.5 First effect of MED configuration

Mass and energy balances provide the exchanged heat, evaporated water at the evaporator and condensed distillate at the pre-heater. Temperature of this first effect is to be parameterized in order to optimize the system. As the temperature must be below 100 °C to avoid scaling onto the evaporator, the pressure in the cell has to be reduced up to the vapor pressure at this temperature.

Equations at the First Effect

Evaporator

$$Q_1 = \dot{m}_g \cdot \left(h_{g\_in\_1} - h_{g\_out\_1}\right) = \dot{m}_{ve1} \cdot (h_{v1} - h_{l1}) \tag{7.44}$$

$$\dot{m}_w = \dot{m}_{re1} + \dot{m}_{ve1}. \tag{7.45}$$

Pre-Heater

$$Q_1 = \dot{m}_{dp1} \cdot (h_{v1} - h_{l1}) = \dot{m}_w \cdot \left(h_{w\_out\_1} - h_{w\_in\_1}\right) \tag{7.46}$$

$$\dot{m}_{ve1} = \dot{m}_{vp1} + \dot{m}_{dp1}. \tag{7.47}$$

Subsequently, a cascade of effects in downward temperature provides heat from the previous effect to the next one to pre-heat the feeding water flow and cooling to condensate the stream of vapor from the previous cell. As previous effect $i - 1$ is warmer than i-effect, the rejected liquid coming from it is superheated, thus it flashes at the operating pressure of the effect resulting in a prior vapor stream. In Fig. 7.6, this concept is shown.

Once the first effect is computed, mass and energy balance equations are applied to flash chamber, evaporator and pre-heater to define the thermodynamic state of

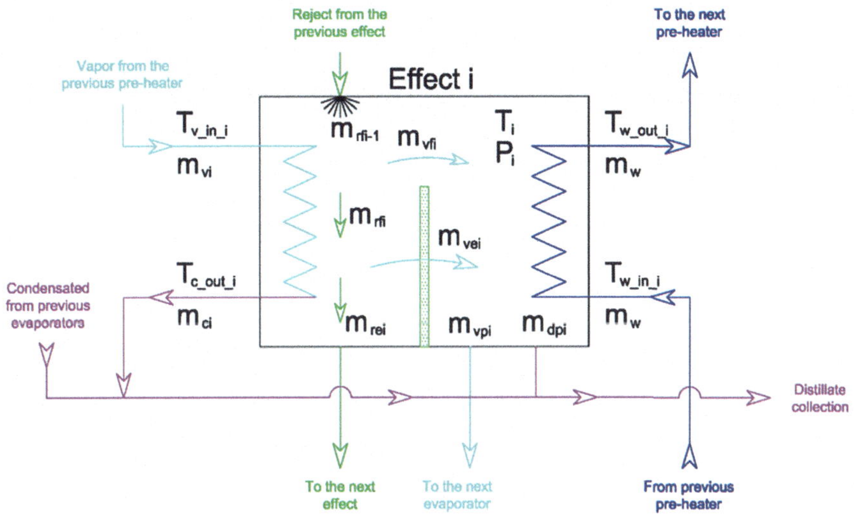

Fig. 7.6 Generic cell of the MED distiller

each flow. The warm flow for the evaporator is taken from the vapor stream leaving the previous pre-heater, $\dot{m}_{vpi-1}$.

Equations at the Generic I-Effect

Flash Chamber

$$\dot{m}_{rfi-1} \cdot h_{rfi-1} = \dot{m}_{vfi} \cdot h_{vfi} + \dot{m}_{rfi} \cdot h_{rfi} \tag{7.48}$$

$$\dot{m}_{rfi-1} = \dot{m}_{vfi} + \dot{m}_{rfi}. \tag{7.49}$$

Evaporator

$$\dot{m}_{vi} = \dot{m}_{vpi-1} \tag{7.50}$$

$$\dot{m}_{vi} = \dot{m}_{ci} \tag{7.51}$$

$$Q_i = \dot{m}_{vi} \cdot \left(h_{v\_\text{in}\_i} - h_{c\_\text{out}\_i}\right) \tag{7.52}$$

$$\dot{m}_{vi} \cdot h_{v\_\text{in}\_i} + \dot{m}_{rfi} \cdot h_{rfi} = \dot{m}_{rei} \cdot h_{rei} + \dot{m}_{vei} \cdot h_{vei} + \dot{m}_{ci} \cdot h_{c\_\text{out}\_i} \tag{7.53}$$

$$\dot{m}_{rfi} = \dot{m}_{rei} + \dot{m}_{vei}. \tag{7.54}$$

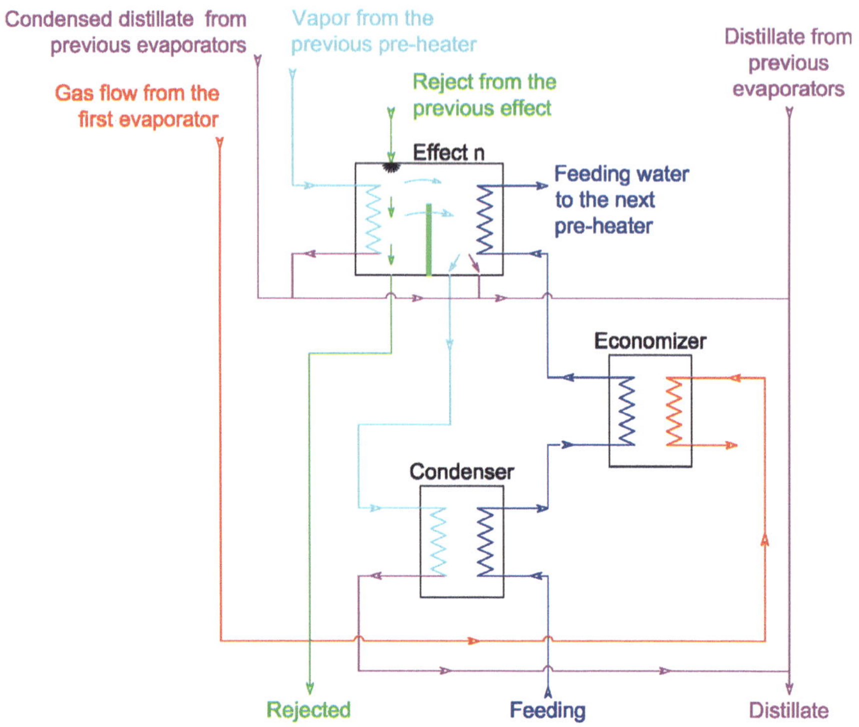

Fig. 7.7 Last effect and feeding configuration for the MED

Pre-heater

$$\dot{m}_{vfi} \cdot h_{vfi} + \dot{m}_{vei} \cdot h_{vei} + \dot{m}_{w} \cdot h_{w\_in\_i} = \dot{m}_{vpi} \cdot h_{vpi} + \dot{m}_{dpi} \cdot h_{dpi} + \dot{m}_{w} \cdot h_{w\_out\_i} \quad (7.55)$$

$$\dot{m}_{vfi} + \dot{m}_{vei} = \dot{m}_{vpi} + \dot{m}_{dpi}. \quad (7.56)$$

Condenser

At last effect, the remaining vapor is sent to the condenser to be condensed by the cool feeding water. Energy delivered at condenser raises the first step temperature of the feeding water.

Downwards, an economizer is previously disposed to the last effect in order to pre-heat the feeding water before going through the sequential pre-heaters of the different effects. The thermal source for this purpose is the residual gases flow from the evaporator at first effect. The configuration can be seen in Fig. 7.7.

To compute the theoretical behavior of the system, the equations have to be solved in an iterative process since the resultant flows are interacting with the inlet flow at the condenser. At this point, there are two unknown flows, so one of them has to be estimated. The feeding water flow is known before entering the condenser. At the

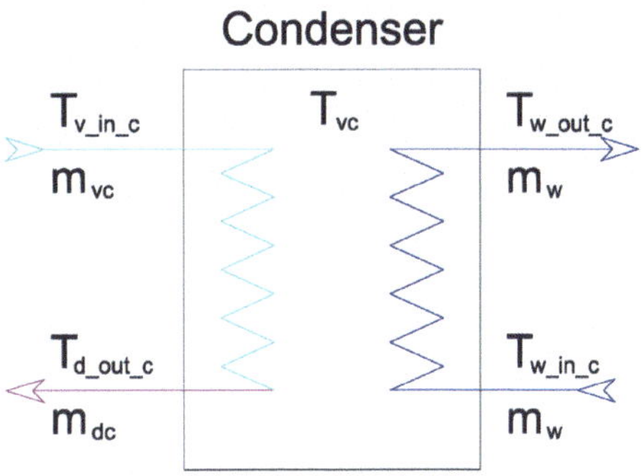

Fig. 7.8 Model of the condenser (*Source* Data compiled by the authors)

condenser outlet, it has to be calculated from the vapor flow from last effect, which is one of the variables to be computed with the system. The temperature of the feeding water at the condenser is estimated. After calculating every effect, it is obtained the value of the vapor from the last effect. This flow applied to the condenser provides a calculated temperature for the feeding water. Correcting it, the process is repeated until both, estimated and calculated temperatures equal.

There are some parameters to be taken into account in order to make the system consistent from the thermodynamic point of view. These are as follows:

- The thermal differential at the economizer is supposed at the gases outlet
- The thermal differential at the evaporator in the first effect at the gases outlet
- The thermal differential in each effect is assumed to be equal.

The system is solved by equaling exchanged heats and assessing mass and energy balances at each effect, condenser and economizer as follows.

Equation at the condenser (Fig. 7.8)

$$Q_c = \dot{m}_w \cdot \left(h_{w\_\text{out}\_c} - h_{w\_\text{in}\_c}\right) = \dot{m}_{v\_c} \cdot h_{v\_\text{in}\_c} - \dot{m}_{d\_c} \cdot h_{d\_\text{out}\_c}. \tag{7.57}$$

Equation at the economizer

$$Q_e = \dot{m}_w \cdot \left(h_{w\_\text{out}\_e} - h_{w\_\text{in}\_e}\right) = \dot{m}_{ge} \cdot \left(h_{ge\_\text{in}\_e} - h_{ge\_\text{out}\_e}\right). \tag{7.58}$$

Inlet conditions are known for both flows, but outlet temperatures are unknown, so an assumption has to be made. The temperature differential at the gas outlet is parameterized in order to relate both outlet temperatures. Under these circumstances, the equation can be solved. Figure 7.9 illustrates it.

Depending on the first effect temperature, the efficiency varies and the higher the temperature, the better the efficiency is. The limit of that is 70 °C since this is the temperature at which scaling might start to occur. As a theoretical exercise, it has been tried the system with different temperatures. Xbre has to be kept under solubility limit to avoid deposition. For the most common salts, this limit is below 50%. Thus, solutions with a greater value are discarded. The second, third and sixth cases present 100% of concentration. It means that only salt remains where all the water has been

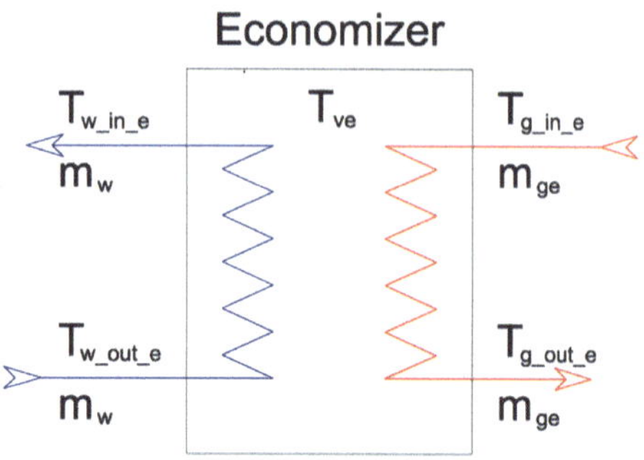

Fig. 7.9 Economizer model

evaporated. Such situation is not acceptable from the operational point of view, since the salt deposition is not removed by the brine flow.

7.3.6.3 Models Coupling

The system is assumed to be run in a continuous mode as a tunnel kiln usually is operated. Thus, it is assumed as well that the MED system runs continuously. Under such circumstances, the system is modeled by coupling both individual models. There are some precedents matching water-energy interaction [15]. The kiln is providing heat permanently to the distiller producing a stream of desalted water. The fraction of water contained within the residual brine has to be added as freshwater as per the factory requirement.

7.3.6.4 MED Consumptions and Outputs

Result Balance of Simple Cascade of Condenser-Evaporator Cells

The result of a simple cascade as per Sect. 7.3.6.1 is shown in Table 7.23.

The distillated water comes from the condensers (m_c), and totalizes 0.1456 kg/s. The efficiency of the system for the plaster mold factory equals:

$$\eta = \frac{m_c}{m_w} = \frac{0.0528 + 0.0928}{1.0498} = 0.1386. \tag{7.59}$$

Regarding the resin mold factory, the efficiency is:

$$\eta = \frac{m_c}{m_w} = \frac{0.0528 + 0.3566}{10.1099} = 0.0405. \tag{7.60}$$

In this case, the amount of water to be treated is much greater. Hence, all the recovered heat is consumed by the distiller and condensed water is notably greater

Table 7.23 First approximation to MED by a simple thermal cascade configuration

Simple cascade of condenser-evaporator cells						
			PMF		RMF	
			Cell 1	Cell 2	Cell 1	Cell 2
Evaporator	m_g	kg/s	2.9677	1.2545	2.9677	4.8455
	T_{g_in}	°C	303	264	303	264
	h_{g_in}	kJ/kg	309	268	309	268
	T_{g_out}	°C	264	100	264	100
	h_{g_out}	kJ/kg	268	101	268	101
	Q_e	kW	120	211	120	809
	T_v	°C	100	100	100	100
	h_l	kJ/kg	419	419	419	419
	h_v	kJ/kg	2,688	2,688	2,688	2,688
	m_v	kg/s	0.0528	0.0928	0.0528	0.3566
Condenser	m_c	kg/s	0.0528	0.0928	0.0528	0.3566
	h_l	kJ/kg	419	419	419	419
	h_v	kJ/kg	2,688	2,688	2,688	2,688
	Q_c	kW	120	211	120	809
	m_w	kg/s	1.0498	1.0498	10.1099	10.1099
	Dh_w	kJ/kg	114	201	12	80
	h_{w_out}	kJ/kg	419	305	197	185
	T_{w_out}	°C	100	73	47	44
	h_{w_in}	kJ/kg	305	105	185	105
	T_{w_in}	°C	73	25	44	25

(0.4093 kg/s). However, efficiency is lower because of the difference of water volumes.

Result Balance for MED

Mass and energy balances for the model applied to both factories yield the following resultant values in Tables 7.24 and 7.25.

Relevant results are summarized in Table 7.26.

In Fig. 6.19, it was stated the proposed scheme to implement the MED distiller. The system combined with MED model has been solved for both types of factory by means of the distiller model. Figure 7.10 shows the resultant balances.

Comparing these balances with Fig. 6.18, water consumption and water residue reductions have been calculated. Results are shown in Table 7.27.

Table 7.24 MED results for a plaster mold factory

Med balance in a plaster mold factory														
Cell		1	2	3	4	5	6	7	8	9	10	11	ECO	CON
T_{cell}	°C	70.00	67.45	64.90	62.35	59.79	57.24	54.69	52.14	49.59	47.04	44.49	41.93	33.76
P_{vapor}	kPas	31.45	28.11	25.07	22.32	19.83	17.59	15.56	13.74	12.11	10.65	9.34	8.17	5.24
P_{vacuum}	kW	3.51	2.86	2.34	1.92	1.57	1.29	1.06	0.87	0.72	0.60	0.50	0.00	0.00
h_w	kJ/kg	293	283	272	261	251	240	229	219	208	197	186	176	142
h_v	kJ/kg	2,632	2,627	2,622	2,617	2,612	2,608	2,603	2,598	2,593	2,589	2,584	2,579	2,564
Flash														
m_{flash}	kg/s	1.0498	0.9478	0.8563	0.7747	0.7024	0.6388	0.5833	0.5354	0.4945	0.4602	0.4319		
m_{stf}	kg/s	0.0000	0.0043	0.0039	0.0035	0.0032	0.0029	0.0026	0.0024	0.0022	0.0021	0.0019		
m_{brf}	kg/s	1.0498	0.9435	0.8524	0.7712	0.6992	0.6359	0.5807	0.5330	0.4923	0.4581	0.4299		
Evaporator														
m_{eva}	kg/s	0.0000	0.0966	0.0861	0.0763	0.0670	0.0583	0.0502	0.0427	0.0356	0.0291	0.0231		
m_{ste}	kg/s	0.1020	0.0871	0.0777	0.0688	0.0604	0.0526	0.0453	0.0385	0.0321	0.0263	0.0208		
m_{bre}	kg/s	0.9478	0.8563	0.7747	0.7024	0.6388	0.5833	0.5354	0.4945	0.4602	0.4319	0.4091		
x_{bre}	%	0.6%	0.6%	0.7%	0.7%	0.8%	0.9%	1.0%	1.1%	1.1%	1.2%	1.3%		
Demister														
m_{std}	kg/s	0.1020	0.0915	0.0816	0.0723	0.0636	0.0555	0.0479	0.0409	0.0344	0.0283	0.0228		
Pre-heater														
m_{wc}	kg/s	1.0498	1.0498	1.0498	1.0498	1.0498	1.0498	1.0498	1.0498	1.0498	1.0498	1.0498		
m_{cop}	kg/s	0.0053	0.0053	0.0053	0.0053	0.0053	0.0053	0.0053	0.0052	0.0052	0.0052	0.0052		
m_{stp}	kg/s	0.0966	0.0861	0.0763	0.0670	0.0583	0.0502	0.0427	0.0356	0.0291	0.0231	0.0176		

(continued)

Table 7.24 (continued)

Med balance in a plaster mold factory

Cell		1	2	3	4	5	6	7	8	9	10	11	ECO	CON
Economizer														
m_{w_eco}	kg/s												1.0498	
m_{rh_eco}	kg/s												1.0942	
$T_{w\_eco\_in}$	°C												33.76	
$T_{rh\_eco\_in}$	°C												80	
$T_{rh\_eco\_out}$	°C												43.76	
Q_{eco_in}	kW												39.93	
Q_{eco_ef}	kW												35.93	
$T_{w\_eco\_out}$	°C												41.93	
Condenser														
m_{dis_con}	kg/s													0.0176
m_{fw_con}	kg/s													1.0498
$T_{fe\_con\_in}$	°C													25.00
$T_{dis\_con\_in}$	°C													44.49
$T_{dis\_con\_out}$	°C													35.00
Q_{eco_in}	kW													42.80
Q_{eco_ef}	kW													38.52
$h_{fe\_con\_in}$	kJ/kg													105
$h_{fe\_con\_out}$	kJ/kg													142
$T_{fe\_con\_out}$	°C													33.77

Table 7.25 MED results for a resin mold factory

Med balance in a resin mold factory													
Cell		1	2	3	4	5	6	7	8	9	10	11	ECO	CON
T_{cell}	°C	70.00	67.45	64.90	62.35	59.79	57.24	54.69	52.14	49.59	47.04	44.49	41.93	33.76
P_{vapor}	kPas	31.45	28.11	25.07	22.32	19.83	17.59	15.56	13.74	12.11	10.65	9.34	8.17	5.24
P_{vacuum}	kW	32.45	26.47	21.67	17.73	14.51	11.89	9.76	8.04	6.65	5.53	4.63	0.00	0.00
h_w	kJ/kg	293	283	272	261	251	240	229	219	208	197	186	176	142
h_v	kJ/kg	2,632	2,627	2,622	2,617	2,612	2,608	2,603	2,598	2,593	2,589	2,584	2,579	2,564
Flash														
m_{flash}	kg/s	9.6997	8.7575	7.9123	7.1584	6.4902	5.9024	5.3897	4.9470	4.5693	4.2519	3.9902		
m_{stf}	kg/s	0.0000	0.0400	0.0361	0.0325	0.0294	0.0267	0.0243	0.0223	0.0205	0.0190	0.0178		
m_{brf}	kg/s	9.6997	8.7175	7.8762	7.1258	6.4608	5.8757	5.3654	4.9247	4.5488	4.2329	3.9724		
Evaporator														
m_{eva}	kg/s	0.0000	0.8929	0.7960	0.7048	0.6192	0.5389	0.4640	0.3941	0.3292	0.2690	0.2134		
m_{ste}	kg/s	0.9422	0.8052	0.7179	0.6356	0.5584	0.4860	0.4184	0.3554	0.2969	0.2426	0.1925		
m_{bre}	kg/s	8.7575	7.9123	7.1584	6.4902	5.9024	5.3897	4.9470	4.5693	4.2519	3.9902	3.7799		
x_{bre}	%	0.6%	0.6%	0.7%	0.7%	0.8%	0.9%	1.0%	1.1%	1.1%	1.2%	1.3%		
Demister														
m_{std}	kg/s	0.9422	0.8452	0.7539	0.6681	0.5878	0.5127	0.4427	0.3777	0.3174	0.2617	0.2103		
Pre-heater														
m_{wc}	kg/s	9.6997	9.6997	9.6997	9.6997	9.6997	9.6997	9.6997	9.6997	9.6997	9.6997	9.6997		
m_{cop}	kg/s	0.0494	0.0492	0.0491	0.0490	0.0489	0.0487	0.0486	0.0485	0.0483	0.0482	0.0481		
m_{stp}	kg/s	0.8929	0.7960	0.7048	0.6192	0.5389	0.4640	0.3941	0.3292	0.2690	0.2134	0.1622		

(continued)

Table 7.25 (continued)

Med balance in a resin mold factory

Cell		1	2	3	4	5	6	7	8	9	10	11	ECO	CON
Economizer														
m_{w_eco}	kg/s												9.6997	
m_{ra_eco}	kg/s												10.1099	
$T_{w\_eco\_in}$	°C												33.76	
$T_{rh\_eco\_in}$	°C												80	
$T_{rh\_eco\_out}$	°C												43.76	
Q_{eco_in}	kW												368.91	
Q_{eco_ef}	kW												332.02	
$T_{w\_eco\_out}$	°C												41.93	
Condenser														
m_{dis_con}	kg/s													0.1622
m_{fw_con}	kg/s													9.6997
$T_{fe\_con\_in}$	°C													25.00
$T_{dis\_con\_in}$	°C													44.49
$T_{dis\_con\_out}$	°C													35.00
Q_{eco_in}	kW													395.43
Q_{eco_ef}	kW													355.89
$h_{fe\_con\_in}$	kJ/kg													105
$h_{fe\_con\_out}$	kJ/kg													142
$T_{fe\_con\_out}$	°C													33.77

Table 7.26 Summary of MED results

MED results		PMF	RMF
Residual water	m^3/year	8,200	75,768
Water saving	m^3/year	13,020	89,521
Distilled water	m^3/year	21,043	194,429
Temperature of recovered heat	°C	288	288
Thermal power required	kW_{th}	321	2,964
Annual thermal energy consumption	kWh/year	1,785,932	16,501,527
$m_{distillated}$	kg/s	0.6407	5.9198
$m_{distillated}$	m^3/year	12,842	118,661
m_{brine}	kg/s	0.4091	3.7799
Electrical power for vacuum	kW	17	159
Annual electrical energy	kWh/year	96,016	887,161
h (kg $_{distillate}$/kg $_{freshwater}$)		0.6103	0.6103
GOR (Grade of recovery = kg $_{distillate}$/kg $_{RH}$)		0.59	0.59

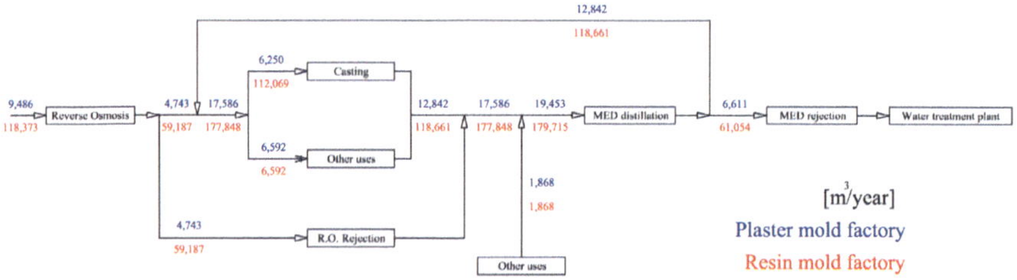

Fig. 7.10 Water saving scheme by MED distiller implementation

Table 7.27 Water consumption and residue reductions by application of a MED distiller

m^3/year	PMF		RMF	
	R.O. Inlet	Water treatment plant	R.O. Inlet	Water treatment plant
Without MED	25,685	20,960	237,322	232,597
With MED	12,665	8,200	147,801	75,768
Reduction	13,020	12,760	89,521	156,829

R.O. Consumptions and Electrical Increases

In the case of the R.O. system, the range of electricity to produce freshwater is around 7,200 kJ/m^3 of freshwater [16] (which is a relative realistic value as per the current state of the art). Under such circumstances, the electrical energy to produce the required amounts of water is respectively:

$$25,685\frac{m^3}{year} \cdot 7,200\frac{kJ}{m^3} \cdot 1\frac{h}{3,600\ s} = 51,370\frac{kWh}{year} \tag{7.61}$$

for plaster mold factory, and

$$237,322\frac{m^3}{year} \cdot 7,200\frac{kJ}{m^3} \cdot 1\frac{h}{3,600\ s} = 474,644\frac{kWh}{year} \tag{7.62}$$

for resin mold factory.

Thus, electrical consumption gets increased because of the required vacuum for the MED distiller. This is the price to recover more than a half of water.

7.3.7 Photovoltaic Plant

The shape of daily demand curves type does not vary much along the months within the period of the year [17]. Nevertheless, the maximum of power depends more on the month. It can be seen a lower level of demand during the months in summer. Daily curves for each month are shown in Fig. 7.11.

Figure 6.11 of Sect. 6.11.1 described the averaged curve of electrical demand in a typical factory of sanitary ware where no saving measures had been carried out and the corresponding adjusted PV plant.

So far some electrical reductions have been achieved, so electrical demand has been cut down as shown in Fig. 7.12.

As a consequence of this fact, the optimized PV plant results smaller. The criteria to select it are the same that there which means to design for the peak of the curve at 1,349 kW. The operation and construction parameters of the resultant plant are shown in Table 7.28.

Corresponding savings are 64.16% of the total required electricity.

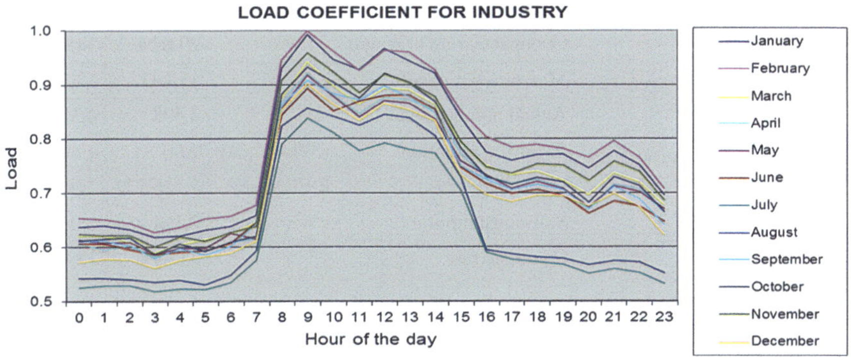

Fig. 7.11 Load coefficient type for an industrial demand curve

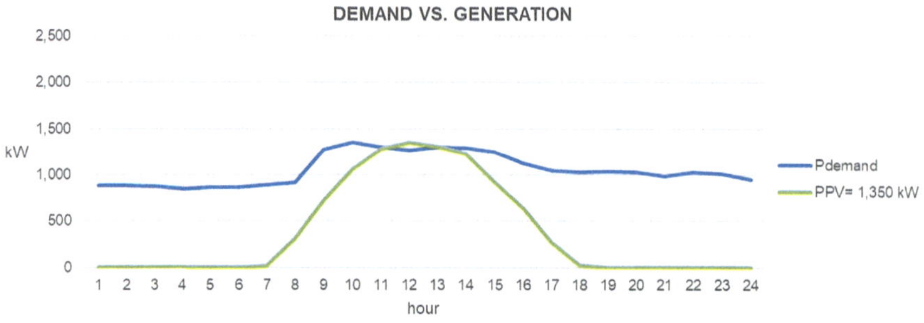

Fig. 7.12 Electrical demand for a factory with saving measures

Table 7.28 Parameters of the optimized PV plant. (*Source* Data compiled by the authors)

Electrical demand		PMF	RMF
Annual energy	kWh/year	5,871,332	5,689,686
$P_{average}$	kW	1,054	1,022
P_{max}	kW	1,349	1,308
Photovoltaic plant		*PMF*	*RMF*
Total power plant	kW	1,349	1,308
Roof surface	m^2	23,100	18,767
Panels length	M	1.956	1.956
Panels width	M	0.992	0.992
Panels surface	m^2	1.940	1.940
Panel power	W	300	300
Panel weigh	kg/panel	26	26
Structure weigh	kg/panel	29	29
Roof resistance	kg/m^2	37	37
Latitude	°	39.81	39.81
Projected length	M	1.503	1.503
Projected surface	m^2	1.490	1.490
Efficiency		17.89%	17.89%
Occupancy rate of panels		29.02%	34.62%
Max number panels		15,499	12,592
Actual number panels		4,498	4,359
Averaged radiation	$kWh/m^2 \cdot day$	4.60	4.60
Generated energy	kWh/year	2,126,277	2,060,495
Self-consumption	kWh/year	2,104,488	2,039,380
Coverage rate		35.84%	35.84%

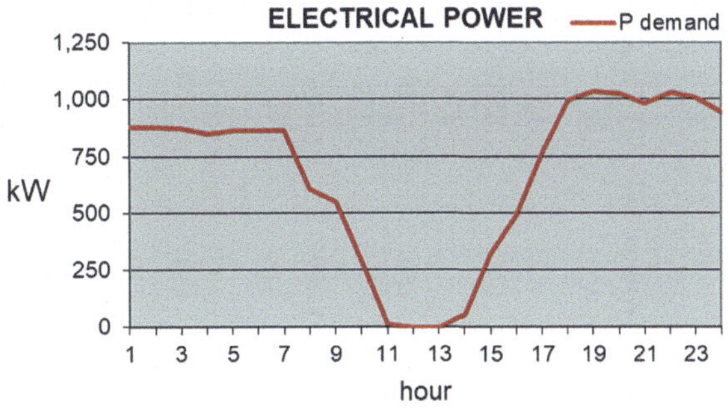

Fig. 7.13 Curve of electrical demand after connecting a photovoltaic plant

7.3.8 Cogeneration [18]

Once the energy map of the factory has been set with energy reductions, this section proposes a heat recovery and a combined heat and power plant implantation to reduce further primary energy by saving fuel in some sub-processes, which are then supplied with heat recovered from kiln and gas turbine as seen in Fig. 6.14 in Sect. 6.11.2.

In the same line of the photovoltaic plant, in the section, it has to be considered that some general reductions of required energy have been achieved, so has the electrical power itself because of the photovoltaic contribution. The electrical demand has been decreased and the shape differs notably due to the impact of solar irradiation distribution. Figure 7.13 shows this new appearance.

As stated in Sect. 6.11.2, the best power design requires total heat recovery to minimize the cost. Under such circumstances, there is an excess of produced electricity during most of the time. Commonly, this power can be released to the network when not required on site deriving economic profits.

Table 7.29 compares the two types of factory, the consumptions before and after implementing the cogeneration plant where the required powers have been reduced by previously described measures. Electrical power of the GT has been set as per supply electricity at any time. When power cannot be sold to the network, the GT has to be regulated off-load to reduce the output.

Distiller and absorption machine have been added as thermal consumers in search of water savings. Nevertheless, it might be desired to look only for just energy to minimize investment. Hence, the power of the plant must be reduced to meet the non-waste restriction. In this sense, a design problem arises from the variability of the electrical demand. The same situation appears when electrical surplus cannot be downloaded to the network. When this situation comes up, the plant has to be designed under different conditions with electrical generated power to match variable requirements. This is the actual difficulty of this section. While the kiln is operated in a steady and continuous way, heat sinks are working with variable cycles. Hence, they have to be planned to work overlapped in such a way that the global heat demand curve is flattened. This means to delay starting time cycles to combine the thermal peak demand of some sinks with the low demand of others. This consideration is equivalent to design with averaged thermal powers. Regarding the turbine, it should

Table 7.29 Power requirement before implementing a combined heat and power plant

		Plaster mold factory				Resin mold factory			
		Without GT	GT $P_{optimal}$	Without GT	GT $P_{optimal}$	Without GT	GT $P_{optimal}$	Without GT	GT $P_{optimal}$
		kWh/year	kWh/year	kW	kW	kWh/year	kWh/year	kW	kW
Electricity	Generated power		6,197,366		1,113		6,288,249		1,129
	Consumed power	3,766,843	3,766,843	677	677	3,650,306	3,650,306	656	656
	Surplus of power		2,430,523		437		2,637,943		474
	Power from network=	3,766,843	0	677	0	3,650,306	0	656	0
Natural gas	GT		20,657,886		3,710		20,960,829		3,765
	Tunnel kiln	7,844,158	5,335,935	1,409	958	7,844,158	7,844,158	1,089	1,089
	Shuttle kiln	2,344,362	1,935,316	421	348	2,344,362	2,344,362	515	515
	Molds dryer	2,004,792		360		0		0	
	Ware dryer	2,112,526		379		2,112,526		414	
	Heating plants	4,351,851		782		609,956		110	
	Boilers	581,856		105		581,856		707	
	Distiller	1,475,110		265		13,629,616		2,448	
	Absorption	4,277,379		768		0		412	
	Wrapping machine	240,000	240,000	100	100	240,000	240,000	100	100
	Canteen	18,560	18,560	10	10	18,560	18,560	10	10
	Thermal savings	4,470,122	4,470,122	803	803	3,820,606	3,820,606	686	686
	Recoverable heat from GT		13,634,205		2,449		13,834,147		2,485
	Recoverable heat from TK	5,064,305	5,064,305	138	138	5,064,305	5,064,305	435	435
	Recoverable heat from SK	907,411	907,411	199	199	907,411	907,411	199	199

(continued)

Table 7.29 (continued)

	Plaster mold factory				Resin mold factory			
	Without GT	GT $P_{optimal}$	Without GT	GT $P_{optimal}$	Without GT	GT $P_{optimal}$	Without GT	GT $P_{optimal}$
	kWh/year	kWh/year	kW	kW	kWh/year	kWh/year	kW	kW
Residual heat	0	4,802,407	0	863	0	2,871,910	0	516
Lack of recovery	8,831,798	0	1,586	0	10,962,237	0	1,969	0
Natural gas from network	14,808,756	4,111,655	3,458	1,537	17,588,712	7,781,441	4,484	1,674
Total Primary Power	27,364,900	4,111,655	5,713	1,537	29,756,397	7,781,441	6,670	1,674

be run at the design point with constant power supplied, but this is not the actual electrical demand of the factory which is varying as per the demand curve so the power has to be adjusted and so the efficiency of the turbine is decreasing as the load does. An hourly demand curve for 1 year is used to simulate the real behavior of the system calculating the variables in every hour according to the real demand and adding them together for the whole year.

Once the configuration and requirements are known, the only question is the size of the turbine. With a small turbine, electricity and natural gas have to be purchased to accomplish the requirements, although it will be run at the design point all the yearlong with the best efficiency. On the other hand, when the turbine is oversized, it will be run off design most of the time, which results in a low-efficiency performance. It has to be optimized as per the real thermal and electrical demands. The optimization can be achieved either from the economical point of view depending on the prices of the energy or from the energy point of view depending on the actual consumption of fuel for the set of gas turbine, thermal requirements within the factory and the power station that supplies the network where the factory is linked to.

The energy optimization only depends on the demand. Thus, the energy optimization will be solved by calculating several pairs of turbine power-annual consumption and then a correlation is calculated by a LSA (Least Square Adjustment).

The economical optimization is not so easy since the optimal power is depending on the two-variable prices. The process to calculate the optimal GT is to find a function, which correlates every price scenario with the corresponding optimal power.

The efficiency of a gas turbine is decreasing with the load [19] as Fig. 7.14 shows:

For this purpose, an artificial neural network is developed, which is trained with several sets of data. Every set includes; NG price, electricity price, optimal power.

Every price scenario is a single optimization problem that has to be solved by iterative minimization of the cost calculated along the year for the pair of prices and the demand curve and yields an optimal turbine power to minimize the total cost of NG and electricity.

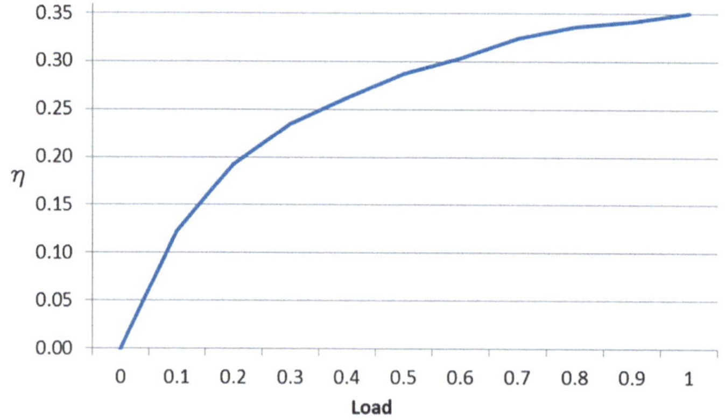

Fig. 7.14 Efficiency of a gas turbine versus load

The ANN is trained with these data to provide the function, which yields the best turbine choice for every pair of prices.

7.3.8.1 Cogeneration Plant Design

The mathematical model for the cogeneration plant is assumed as a conventional gas turbine process as per Fig. 7.15. Air is compressed before entering the combustion chamber to be heated by natural gas combustion. The expansion of the flue gases throughout the turbine produces the electrical output and residual heat.

The thermodynamic process is shown in Fig. 7.16.

The red line defines the real process whose stages are described as follows:

- 0–2: Adiabatic compression from ambient P_0 and T_0 up to P_2 and T_2 defined by compression rate of the GT.
- 2–3: Isobaric combustion from natural gas
- 3–5: Adiabatic expansion from ambient P_3 and T_3 up to P_5 and T_5 defined by ambient pressure.
- 5–0: Cooling of residual gas up to ambient T_0.

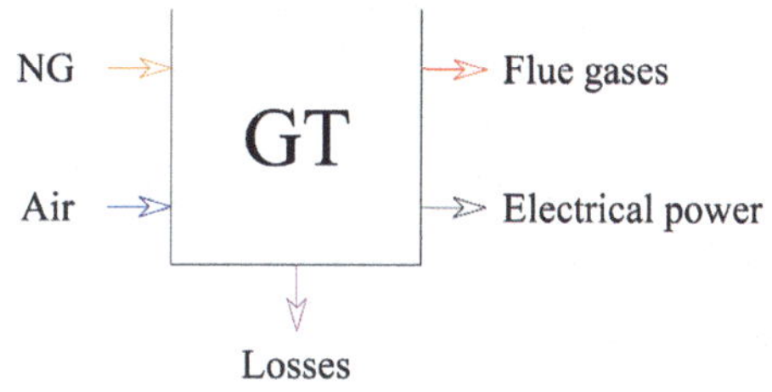

Fig. 7.15 Scheme of GT cogeneration model

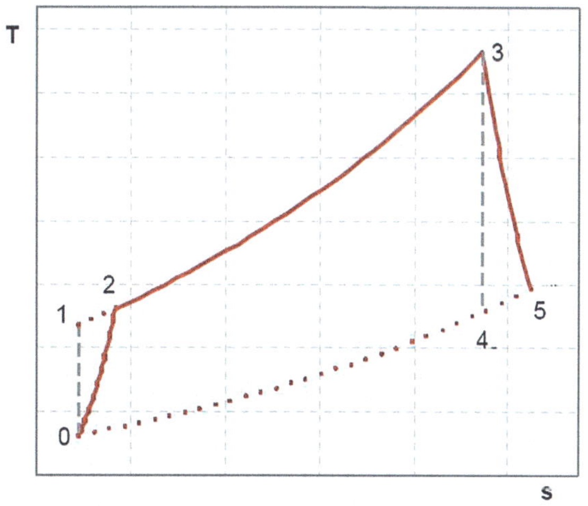

Fig. 7.16 Scheme of thermodynamic process of the gas turbine

Corresponding equations are as follows:

Compression:

$$P_2 = r_c \cdot P_0 \tag{7.63}$$

$$W_{\text{comp}} = m_a \cdot (h_2 - h_0) \tag{7.64}$$

$$T_2 = T_0 \cdot \left(\frac{P_2}{P_0}\right)^{\frac{\gamma-1}{\gamma}} \tag{7.65}$$

$$\eta_{\text{comp}} = \frac{h_1 - h_0}{h_2 - h_0} \tag{7.66}$$

where:

P: pressure (Pa)

T: Temperature (°C)

W_{comp}: Compression power (kW)

m_a: Air mass flow(kg/s)

h: specific enthalpy (kJ/kg)

γ: Adiabatic constant of the air (Dimensionless)

η_{comp}: Isentropic efficiency of the compressor (Dimensionless).

Combustion:

$$m_a \cdot h_2 + m_{NG} \cdot LHV = m_g \cdot h_3 \tag{7.67}$$

$$m_a + m_{NG} = m_g \tag{7.68}$$

where:

m_a: air mass flow (kg/s)

m_{NG}: natural gas mass flow (kg/s)

m_g: flue gas mass flow (kg/s)

h: specific enthalpy (kJ/kg)

LHV: low heat value of the fuel (kJ/kg).

GT expansion:

$$W_{GT} = m_g \cdot (h_3 - h_5) \tag{7.69}$$

$$T_5 = T_3 \cdot \left(\frac{P_5}{P_3}\right)^{\frac{\gamma-1}{\gamma}} \tag{7.70}$$

$$\eta_{GT} = \frac{h_5 - h_3}{h_4 - h_3} \tag{7.71}$$

where:

W_{GT}: GT power (kW)

η_{GT}: Isentropic efficiency of the GT (Dimensionless).

Conventional Factory Consumption

Primary energy is calculated under the following hypothesis.

- An efficiency of 0.35 is considered in the case of the electricity from primary energy to electrical energy [20]
- The gas consumption is assumed to be primary energy

Once the available powers are quantified, the annual cost and primary energy have to be calculated as follows:

For the conventional factory, the cost is calculated by multiplying each consumption by the corresponding price in every hour, and summing them all together for the whole year:

$$C_{cf} = \sum_{i=1}^{8,760} \left(p_{ng} \cdot (Q_{TK} + Q_{SK} + Q_{MD} + Q_{BO} + Q_{HVU} + Q_{WD}) + p_{\text{elec}} \cdot N_{\text{eleci}}\right). \tag{7.72}$$

When tried different price scenarios, the powers do not change but the cost is varying as per the prices variation. However, primary energy is the same for any economic scenario.

On the other hand, the primary energy is calculated in a similar way. Thermal power for every consumer is added and then the electrical power is divided by the efficiency of the electrical network (assumed equal to 0.35 [20]).

$$W_{cf} = \sum_{i=1}^{8,760} \left(Q_{TK} + Q_{SK} + Q_{MD} + Q_{BO} + Q_{HVU} + Q_{WD} + \frac{N_{\text{eleci}}}{0.35}\right). \tag{7.73}$$

Results of the system for a conventional factory where no heat recovery has been applied are shown in Table 7.30 (Fig. 7.17).

Table 7.30 Cost and primary energy for a CF

C_{cf} (€/year)		p_{elec} (€/kWh)					W_{cf}
		0.08	0.09	0.1	0.11	0.12	kWh/year
p_{NG} (€/kWh)	0.023	826,008	863,744	901,480	939,216	976,952	33,569,522
	0.024	848,795	886,531	924,267	962,003	999,739	
	0.025	871,583	909,319	947,055	984,791	1,022,527	
	0.026	894,371	932,107	969,843	1,007,579	1,045,315	
	0.027	917,159	954,895	992,631	1,030,367	1,068,103	

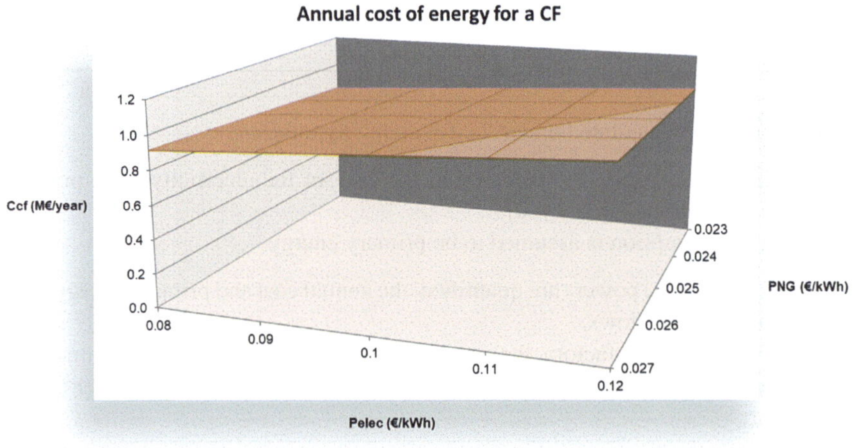

Fig. 7.17 Energy cost as a bi-variant function of fuel and electricity prices (*Source* Data compiled by the authors)

Water-Saving Measures

When water-saving measures are carried out to improve environmental efficiency, demand for heat increases up to that point that a cogeneration plant is convenient in order to profit residual heat.

- The water saving is transformed into primary energy as the thermal energy required for the MED to distillate the needed quantity. MED efficiency is assumed to be $\eta_{MED} = 2.4 \cdot 10^5 \frac{\text{kJ}}{\text{m}^3}$ [13].
- In the case of the absorption machine, the primary energy is calculated through the COP (coefficient of performance).

Thermal power to distillate water is calculated by:

$$Q_{MED} = \dot{m}_w \cdot \eta_{MED}. \tag{7.74}$$

The thermal power to condense water is:

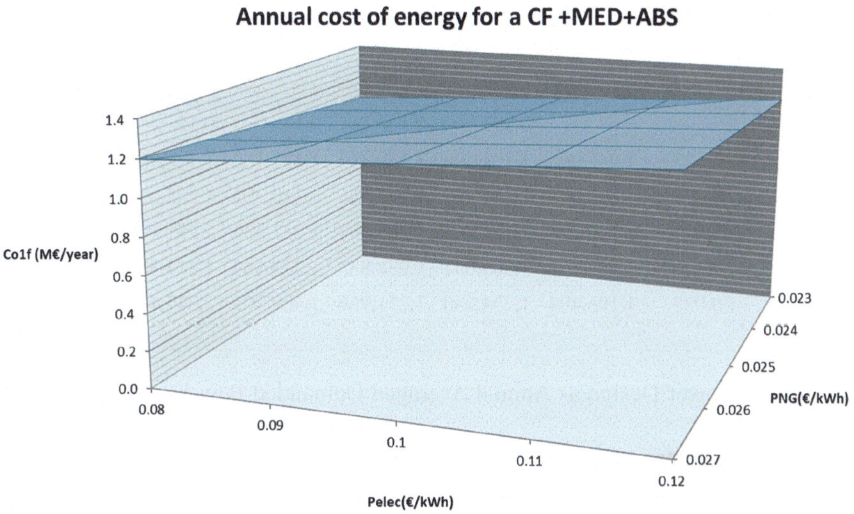

Fig. 7.18 Energy cost as a bi-variant function of fuel and electricity prices (*Source* Data compiled by the authors)

$$Q_{ABS} = \frac{\dot{m}_w \cdot \left(h_{\text{vapor}} - h_{\text{liquid}}\right)}{COP_{ABS}}. \tag{7.75}$$

With this new thermal power, the equations to calculate the increased thermal energy derived from the water saving would be:

$$C_{o1f} = \sum_{i=1}^{8,760} \begin{pmatrix} p_{ng} \cdot (Q_{TK} + Q_{SK} + Q_{MD} + Q_{BO} + Q_{HVU} + Q_{WD} + Q_{MED} + Q_{ABS}) \\ + p_{\text{elec}} \cdot N_{\text{eleci}} \end{pmatrix} \tag{7.76}$$

$$W_{o1f} = \sum_{i=1}^{8,760} \begin{pmatrix} Q_{TK} + Q_{SK} + Q_{MD} + Q_{BO} + Q_{HVU} + Q_{WD} + Q_{MED} + Q_{ABS} \\ + \frac{N_{\text{eleci}}}{0.35} \end{pmatrix}. \tag{7.77}$$

Results for a CF improved with MED distiller and an absorption machine are shown (Fig. 7.18).

Comparing values of Tables 7.30 and 7.31, it can be seen the resources cost for water saving. Primary energy increases by 36.17% and economic cost increase ranges from 24 to 31% depending on the energy price scenario.

Table 7.31 Cost and primary energy for a CF improved with a MED distiller and an absorption machine

C_{olf} (€/year)		p_{elec} (€/kWh)					W_{olf}
		0.08	0.09	0.1	0.11	0.12	kWh/year
p_{NG} (€/kWh)	0.023	1,063,960	1,101,696	1,139,432	1,177,168	1,214,904	43,915,283
	0.024	1,097,094	1,134,830	1,172,566	1,210,302	1,248,038	
	0.025	1,130,227	1,167,963	1,205,699	1,243,435	1,281,171	
	0.026	1,163,361	1,201,097	1,238,833	1,276,569	1,314,305	
	0.027	1,196,494	1,234,230	1,271,966	1,309,702	1,347,438	

Gas Turbine Power Design as Annual Averaged Demanded Power

In order to reduce primary energy, the cogeneration plant is applied to optimize total energy consumption.

Considering the curve demand as per Fig. 7.13, cost and primary energy should be calculated as follows:

Load degree is calculated with the demand curve of electrical power and the gas electrical turbine power

$$Load = \frac{N_{\text{elec}}}{N_{GT}}. \tag{7.78}$$

When load is less than 1, the turbine runs off design with a lower efficiency but all the electricity is supplied by it. When the load is greater than 1, the turbine is run on the design point but electricity above the turbine power has to be purchased from electrical network.

The actual efficiency of the turbine under off-load operation is obtained from Fig. 7.14.

For the periods where $N_{GT} < N_{\text{elec}}$, the difference has to be supplied from network:

$$N_{\text{network}} = N_{\text{elec}} - N_{GT}. \tag{7.79}$$

Then it can be calculated the gas consumption of the turbine:

$$G_{GT} = \frac{N_{GT}}{\eta_{GT}}. \tag{7.80}$$

The exhausting heat of the turbine is calculated by means of the heat-ratio of the turbine

$$Q_{GT} = \text{heat}_{\text{ratio}} \cdot N_{GT}. \tag{7.81}$$

Table 7.32 Cost and primary energy for an optimized factory with a GT sized by averaged electrical power

C_{o2f} (€/year)		p_{elec} (€/kWh)					W_{o2f}
		0.08	0.09	0.1	0.11	0.12	kWh/year
p_{NG} (€/kWh)	0.023	645,042	653,715	662,388	671,062	679,735	27,506,570
	0.024	670,070	678,743	687,417	696,090	704,763	
	0.025	695,099	703,772	712,445	721,118	729,792	
	0.026	720,127	728,800	737,474	746,147	754,820	
	0.027	745,155	753,829	762,502	771,175	779,849	

With all the powers calculated already, they can be obtained cost and energy

$$C_{o2f} = \sum_{i=1}^{8,760} \left(p_{ng} \cdot (Q_{\text{demand}} - Q_{RH}) + p_{\text{elec}} \cdot N_{\text{network}i}\right) \tag{7.82}$$

$$W_{o2f} = \sum_{i=1}^{8,760} \left(Q_{\text{demand}} - Q_{RH} + \frac{N_{\text{network}i}}{0.35}\right) \tag{7.83}$$

where:

C_{o2f}: annual cost of energy

p_{ng}: price of natural gas

Q_{demand}: thermal load demand

Q_{RH}: recovered heat

p_{elec}: price of electricity

N_{network}: electricity supplied by the local network

W_{o2f}: required primary energy.

Results for a factory improved with a GT are shown in Table 7.32 (Fig. 7.19).

Comparing values of Tables 7.31 and 7.32, they can be seen the energy and cost savings when applying a cogeneration plant. Primary energy decreases by 39.83% and economic cost decrease ranges from 29 to 43% depending on the energy prices scenario.

Energy Optimization. The Power of the Optimal Gas Turbine to Minimize the Primary Energy

In the previous section, it has been considered that turbine power was known as the electrical average of the year. However, this is not the real behavior, so electrical power of the GT has to be adjusted for either the lowest energy or cost. The

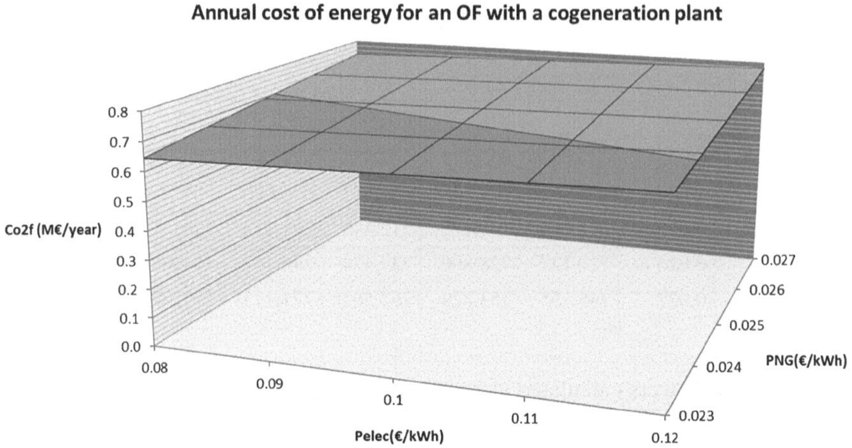

Fig. 7.19 Energy cost as a bi-variant function of fuel and electricity prices (*Source* Data compiled by the authors)

process to optimize the energy involves an iterative procedure by a GRG method. The authors have tried to define a LSA polynomial function of GT power that yields the primary energy whose derivative provides the direction of the gradient. The function was set by a LSA where the pairs are found by solving the set of equations in the previous section for each supposed power providing the correspondent consumption of primary energy. The polynomial function was:

$$W_{o3f} = a_0 + a_1 \cdot N_{GT} + a_2 \cdot N_{GT}^2 + a_3 \cdot N_{GT}^3 + a_4 \cdot N_{GT}^4 + a_5 \cdot N_{GT}^5 \quad (7.84)$$

and its derivative function:

$$\frac{dW_{o3f}}{dN_{GT}} = 0 = a_1 + 2 \cdot a_2 \cdot N_{GT} + 3 \cdot a_3 \cdot N_{GT}^2 + 4 \cdot a_4 \cdot N_{GT}^3 + 5 \cdot a_5 \cdot N_{GT}^4. \quad (7.85)$$

Polynomial coefficients have result:

a0=	2.8117E+07
a1=	−2.1547E+04
a2=	8.9313E+01
a3=	−2.1226E-01
a4=	2.5328E-04
a5=	−1.1521E-07

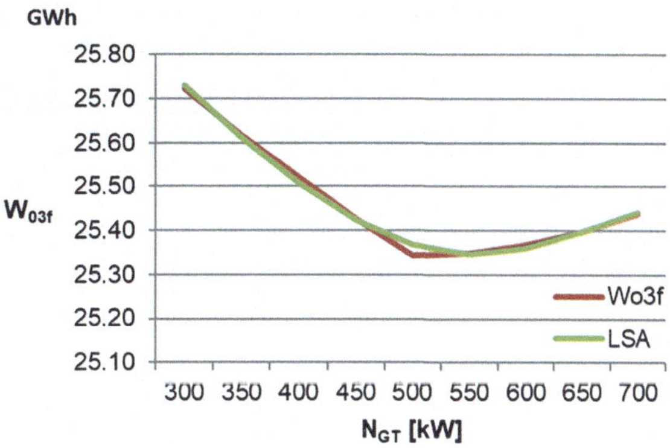

Fig. 7.20 Consumed primary energy as a correlated function of the power of the gas turbine

Table 7.33 Comparison of LSA and gradient descent correlations

Minimum of the primary energy			
	by LSA	by GRG	
$N_{\text{elec } GT \text{ optimal}}=$	556	515	kW
$DW_{of}/DN_{\text{elec } GT}=$	0.00E+00		
$C_{o3fminimum}=$	601,704	603,728	€/year
$C_{o3fmaximum}=$	740,339	746,870	€/year
$W_{o3f}=$	25,351,762	25,340,811	kWh/year

The corresponding function is represented in Fig. 7.20.

Nevertheless, this approximation resulted not good enough to find the minimum of the actual function. Thus, the procedure to find the minimum of the primary energy has been set as a gradient descent method.

Compared resultant values are shown in Table 7.33.

Once the optimal power design is set, cost and primary energy can be calculated by:

$$C_{o3f} = \sum_{i=1}^{8,760} \left(p_{ng} \cdot (G_{GT} + Q_{\text{tunnel kiln}} + Q_{\text{shuttle kiln}} + Q_{\text{demand}} - Q_{\text{RH}}) + p_{\text{elec}} \cdot N_{\text{network}i}\right) \tag{7.86}$$

$$W_{o3f} = \sum_{i=1}^{8,760} \left(G_{GT} + Q_{\text{tunnel kiln}} + Q_{\text{shuttle kiln}} + Q_{\text{demand}} - Q_{\text{RH}} + \frac{N_{\text{network}i}}{0.35}\right). \tag{7.87}$$

For the purpose of this case study, the gradient descent method has been selected and resultant values of cost and primary energy are shown in Fig. 7.21.

Comparing values of Tables 7.32 and 7.34, they can be seen the energy and cost savings when optimizing the gas turbine of the cogeneration plant to minimize the

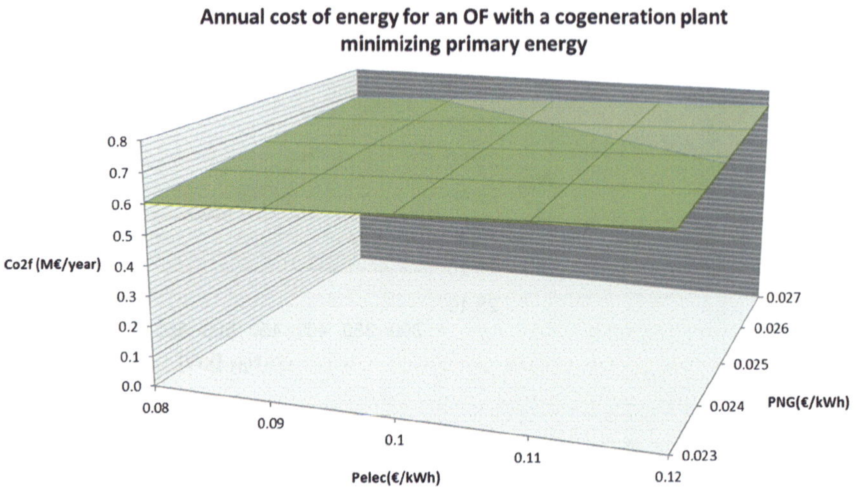

Fig. 7.21 Energy cost as a bi-variant function of fuel and electricity prices (*Source* Data compiled by the authors)

Table 7.34 Primary energy and cost values corresponding to the optimal GT

C_{o3f} (€/year)		p_{elec} (€/kWh)					W_{o3f}
		0.08	0.09	0.1	0.11	0.12	kWh/year
p_{NG} (€/kWh)	0.023	603,728	618,350	632,973	647,595	662,218	25,340,811
	0.024	624,891	639,513	654,136	668,758	683,381	
	0.025	646,054	660,676	675,299	689,921	704,544	
	0.026	667,217	681,839	696,462	711,084	725,707	
	0.027	688,380	703,002	717,625	732,247	746,870	

primary energy. Primary energy decreases by 7.87% and economic cost decrease ranges from 3 to 8% depending on the energy prices scenario.

Economical Optimization. the Power of the Optimal Gas Turbine to Minimize the Annual Cost as a Function of the Prices of the Energy

The optimization process from the economical point of view depends on two variables, so it is not possible to carry it out by a LSA. The process is led by a gradient descent method where the bivariate function is set by an ANN according with the scheme of Fig. 7.22.

Correlation function is defined as follows:

$$VO(i) = \ln(a(i))(i = 1, 2) \tag{7.88}$$

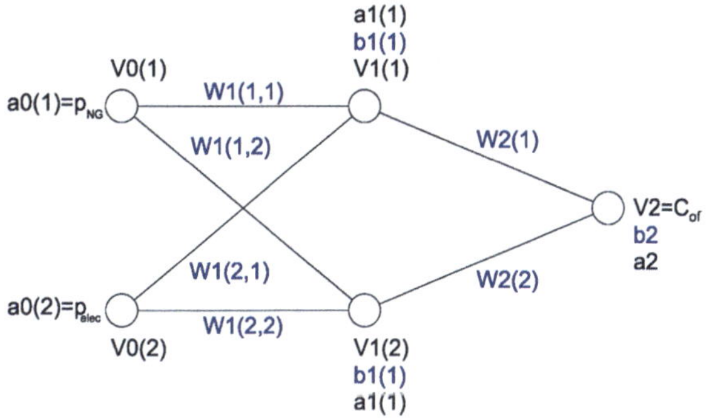

Fig. 7.22 Scheme of the ANN correlation for cost = f(P_{NG})

$$a1(j) = b1(j) + \sum_{i=1}^{2} a0(i, j) \cdot W1(i, j)(j = 1, 2) \tag{7.89}$$

$$V1(j) = e^{a1(j)}(j = 1, 2) \tag{7.90}$$

$$\begin{aligned} f(p_{NGn}, p_{\text{elecn}}) = a2 = V2 = b2 + \sum_{j=1}^{2} V1(j) \cdot W2(j) = b2 \\ + \sum_{j=1}^{2} W2(j) \cdot e^{b1(j)} \cdot \prod_{i=1}^{2} a0(i)^{W1(i,j)}. \end{aligned} \tag{7.91}$$

The ANN is defined by the parameters in blue: $b1(j)$, $W1(i, j)$, $b2$ and $W2(j)$, which have to be iteratively approximated to the optimal solution where the error between correlated output and real one is less than the fixed tolerance. At this point, the ANN would yield a cost for every pair of energy prices as inputs

Since $a0(1)$ and $a0(2)$ correspond respectively with P_{NG} and P_{elec}, the correlated function is:

$$\begin{aligned} f(p_{NGn}, p_{\text{elecn}}) = b2 + W2(1) \cdot e^{b1(1)} \cdot P_{NG}^{W1(1,1)} \cdot P_{\text{elec}}^{W1(2,1)} \\ + W2(2) \cdot e^{b1(2)} \cdot P_{NG}^{W1(1,2)} \cdot P_{\text{elec}}^{W1(2,2)}. \end{aligned} \tag{7.92}$$

Energy is calculated for each optimized GT power as described in 0.

The process is carried out with a set of "N" known training values where there are two inputs defined for energy prices and the actual corresponding output for the annual total cost, energy or the optimal GT power.

$$N_{\text{elec}} = f_1(p_{NGn}, p_{\text{elecn}}) \tag{7.93}$$

$$C_{o4fn} - f_2(p_{NGn}, p_{\text{elecn}}) \tag{7.94}$$

Table 7.35 Optimal power of the GT to minimize annual cost as f(p_{elec}, p_{NG}). (*Source* Data compiled by the authors)

$N_{elec\ GT}$ (kW)		p_{elec} (€/kWh)				
		0.08	0.09	0.1	0.11	0.12
p_{NG} (€/kWh)	0.023	867	881	948	987	999
	0.024	867	877	948	987	999
	0.025	774	867	882	987	999
	0.026	682	867	881	948	987
	0.027	586	867	877	882	987

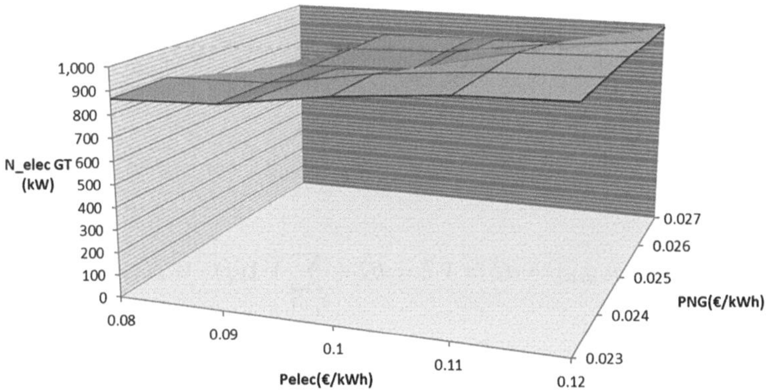

Fig. 7.23 Optimal power of the GT to minimize annual cost as f(p_{elec}, p_{NG}) (*Source* Data compiled by the authors)

$$W_{o4fn} = f_3(p_{NGn}, p_{elecn}). \tag{7.95}$$

This set of training values is got by solving "N" times the equations system in Section "Energy Optimization. the Power of the Optimal Gas Turbine to Minimize the Primary Energy" for different combinations of natural gas and electricity prices. For every pair of energy price, it is calculated the optimal GT power to minimize the annual cost. Thus, this cost and corresponding energy are obtained for this optimal power. Due to the complexity of the system, it cannot be directly solved. Every training value has to be calculated by another iterative process assuming a starting optimal power for the stated prices and then adjusting it to the best solution of gas turbine size by minimizing the cost by a gradient descent method (Table 7.35 and Fig. 7.23).

Corresponding annual cost and consumed primary energy for each optimal power are shown in the following tables and figures (Tables 7.36, 7.37, Figs. 7.24 and 7.25).

The ANN is then trained with these sets of values up to minimize the error of difference between the correlated cost, primary energy or electrical power and the

Table 7.36 Minimized annual cost as f(p_{elec}, p_{NG}). (*Source* Data compiled by the authors)

C_{o4f} (€/year)		p_{elec} (€/kWh)				
		0.08	0.09	0.1	0.11	0.12
p_{NG} (€/kWh)	0.023	592,968	595,133	596,573	596,985	597,239
	0.024	617,937	620,212	622,103	622,752	623,075
	0.025	642,543	645,240	647,321	648,519	648,912
	0.026	665,816	670,210	672,409	674,102	674,681
	0.027	688,192	695,179	697,474	699,508	700,448

Table 7.37 Corresponding energy for the minimized annual cost as f(p_{elec}, p_{NG}). (*Source* Data compiled by the authors)

W_{o4f} (kWh/year)		p_{elec} (€/kWh)				
		0.08	0.09	0.1	0.11	0.12
p_{NG} (€/kWh)	0.023	25,636,216	25,662,507	25,798,225	25,879,694	25,879,694
	0.024	25,636,095	25,653,411	25,798,224	25,879,693	25,907,888
	0.025	25,519,736	25,636,315	25,664,006	25,879,693	25,907,886
	0.026	25,421,265	25,636,215	25,662,505	25,798,223	25,879,694
	0.027	25,362,491	25,636,095	25,653,411	25,664,007	25,879,693

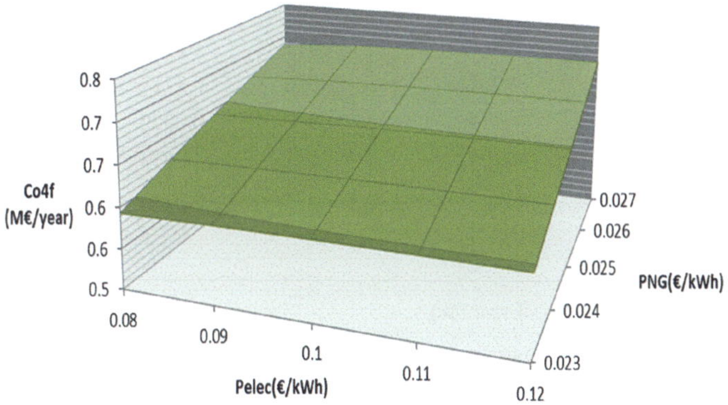

Fig. 7.24 Minimized annual cost as f(p_{elec}, p_{NG}) (*Source* Data compiled by the authors)

actual one. The resultant coefficients and error for the three different correlations are shown in Table 7.38.

The maximal and averaged error of correlation with these ANN's results (Table 7.39).

The correlation for electrical power of the GT results not good enough. However, it is quite accurate for energy and cost. It may be needed to try another ANN structure, but this is out of the scope of this book.

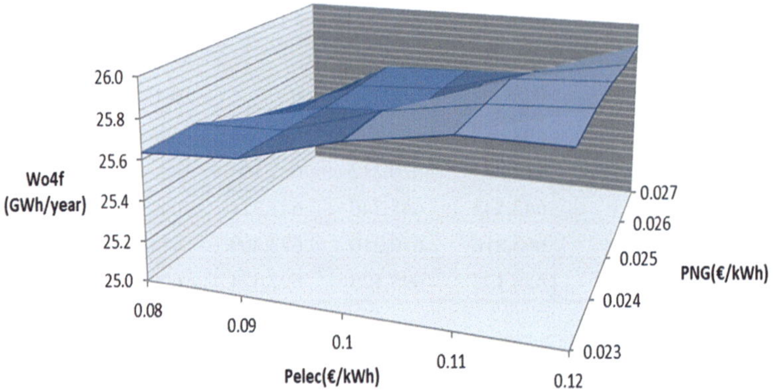

Fig. 7.25 Corresponding energy for the minimized annual cost as f(p_{elec}, p_{NG}) (*Source* Data compiled by the authors)

Table 7.38 ANN coefficients for cost, primary energy and optimal electrical GT power correlations

	N_{elec}	C	W
	kW	€/year	kWh/year
$W1(1,1)=$	0.3828	0.0000	20,031,982
$W1(1,2)=$	0.0000	0.9514	0
$W1(2,1)=$	0.8723	0.0000	0
$W1(2,2)=$	0.6433	0.0261	25,366,577
$W2(1)=$	3.2675	1.5751	0
$W2(2)=$	4.9676	6.3717	13,137,439
$b1(1)=$	0.0000	0.0000	0
$b1(2)=$	6.6431	15.0974	4,483,615
$b2=$	32.4472	7.0115	25,717,315

Table 7.39 Maximum and averaged error for ANN's correlation. (*Source* Data compiled by the authors)

Error (%)			
	N_{elec} (%)	C (%)	W (%)
Maximum	33.69	0.49	1.40
Average	5.12	0.17	0.50

7.4 Summary of Optimization Results

Once the results for the different configuration have been achieved, it can be concluded that:

- Water savings produce and impact in energy consumption and therefore in the corresponding cost.

Table 7.40 Results of the different configuration for the cogeneration plant

	N_{elec}	C_{min}	C_{max}	W_{min}	W_{max}
	kW	€/year	€/year	kWh/year	kWh/year
CF	–	826,008	1,068,103	33,569,522	
O1F	–	1,063,960	1,347,438	43,915,285	
O2F	678	645,042	779,849	27,506,570	
O3F	515	603,728	746,870	25,340,811	
O4F	$f(p_{NG}, p_{elec})$	592,968	700,448	592,968	700,448

- Introducing cogeneration concept to recover heat reduces primary energy consumption and the corresponding cost.
- The size of this cogeneration plant defines the global consumption which can be optimized for either, minimal energy or cost.

Table 7.40 shows these quantified conclusions.

References

1. van Vugt M, Moye A, Sivakumar S (2019) Computational modelling approaches to meditation research: why should we care? Curr Opin Psychol 28:49–53. https://doi.org/10.1016/j.copsyc.2018.10.011
2. Brailsford SC, Eldabi T, Kunc M, Mustafee N, Osorio AF (2019) Hybrid simulation modelling in operational research: a state-of-the-art review. Eur J Oper Res 278(3):721–737. https://doi.org/10.1016/j.ejor.2018.10.025
3. Lamine CM, Said Z (2014) Energy analysis of single effect absorption chiller (LiBr/H2O) in an Industrial manufacturing of detergent. Energy Proc 50:105–112. https://doi.org/10.1016/j.egypro.2014.06.013
4. Various (2014) Guide to meteorological instruments and methods of observation, 8th edn. World Meteorological Organization
5. Nannarone A, Toro C, Sciubba E (2018) Multi-effect distillation desalination process: modeling and simulation. In: Presented at the 30th international conference on efficiency, cost, optimization, simulation and environmental impact of energy systems, San Diego, USA, July 2017. https://www.researchgate.net/publication/318207357_Multi-Effect_Distillation_Desalination_Process_Modeling_and_Simulation. Accessed 05 Nov 2018
6. Frantz C, Seifert B (2015) Thermal analysis of a multi effect distillation plant powered by a solar tower plant - ScienceDirect 69:1928–1937. https://doi.org/10.1016/j.egypro.2015.03.190
7. Ophir A, Gendel A (1994) Adaptation of the multi-effect distillation (MED) process to yield high purity distillate for utilities, refineries and chemical industry. Desalination 98(1):383–390. https://doi.org/10.1016/0011-9164(94)00164-2
8. El Ebaidi SK (2013) Water chemical analysis for different stages in Tobruk desalination plant. In: Presented at the the 13th international conference on materials science and its applications in oil and gas industries, Benghazi, August 2013
9. Chernozubov VB, Zaostrovsky FP, Shatsillo VG, Golub SI, Novilov EP, Tkach VI (1966) Prevention of scale formation in distillation desalination plants by means of seeding. Desalination 1(1):50–60. https://doi.org/10.1016/S0011-9164(00)84007-2

10. Cuviella-Suárez C, Colmenar-Santos A, Borge-Diez D, López-Rey Á (2018) Management tool to optimize energy and water consumption in the sanitary-ware industry. J Clean Prod 197, Part 1:280–296. https://doi.org/10.1016/j.jclepro.2018.06.195
11. Ammar Y, Li H, Walsh C, Thornley P, Sharifi V, Roskilly AP (2013) Reprint of 'Desalination using low grade heat in the process industry: challenges and perspectives'. Appl Therm Eng 53(2):234–245. https://doi.org/10.1016/j.applthermaleng.2012.11.010
12. Dastgerdi HR, Whittaker PB, Chua HT (2016) New MED based desalination process for low grade waste heat. Desalination 395:57–71. https://doi.org/10.1016/j.desal.2016.05.022
13. Hatzikioseyian A, Kousi P (2003) Modelling and thermodynamic analysis of a Multi-effect Distillation (MED) plant for seawater desalination. ResearchGate, January 2003. https://www.researchgate.net/publication/242541781_MODELLING_AND_THERMODYNAMIC_ANALYSIS_OF_A_MULTI_EFFECT_DISTILLATION_MED_PLANT_FOR_SEAWATER_DESALINATION. Accessed 24 Jan 2017
14. Chen Q, Ja MK, Li Y, Chua KJ (2018) Experimental and mathematical study of the spray flash evaporation phenomena. Appl Therm Eng 130:598–610. https://doi.org/10.1016/j.applthermaleng.2017.11.018
15. Dubreuil A, Assoumou E, Bouckaert S, Selosse S, Maı¨zi N (2013) Water modeling in an energy optimization framework – the water-scarce middle east context. Appl Energy 101:268–279. https://doi.org/10.1016/j.apenergy.2012.06.032
16. Fritzmann C, Löwenberg J, Wintgens T, Melin T (2007) State-of-the-art of reverse osmosis desalination. Desalination 216(1):1–76. https://doi.org/10.1016/j.desal.2006.12.009
17. Cuviella-Suárez C, Colmenar-Santos A, Castro-Gil M (2012) Tri-generation system to couple production to demand in a combined cycle. Energy 40(1):271–290. https://doi.org/10.1016/j.energy.2012.01.073
18. Cuviella-Suárez C, Colmenar-Santos A, Borge-Diez D, Rosales-Asensio E (2019) Sanitary-ware factories: heat recovery strategies to optimize energy and water consumption. Energy Proc. 157:719–736. https://doi.org/10.1016/j.egypro.2018.11.238
19. Lazzaretto A, Andrea T (2001) Analytical and neural network models for gas turbine design and off-design simulation. Int J Thermodyn 4:173–182. https://doi.org/10.5541/ijot.78
20. Lecomte T (2010) Best Available Techniques (BAT) Reference Document for Large Combustion Plants - Industrial Emissions Directive 2010/75/EU (Integrated Pollution Prevention and Control), p 986

Chapter 8
Energy Supply Versus Energy Demand

8.1 Variability of the Demand

In Sect. 6.11.2, a cogeneration plant was proposed in order to reduce the primary energy consumption by recovering exhausting gases towards other thermal consumers. Nevertheless, it is not so easy to size related equipment due to the variability of the demand. The way the machinery demands energy depends on multiple factors;

- heating and ventilation units are led by set points of temperature and humidity within the casting hall
- ambient air conditions are quite different along the year for both, temperature and humidity
- solar irradiation varies from zero to maximal values on a daily basis
- psychrometric condition of a ware dryer are very different in just a 10-hour cycle
- heating power for a shuttle kiln ranges 0–1,200 °C in a 15-hour cycle.

As an example given, Fig. 8.1 shows the variable thermal consumers shapes of the demand.

On the other hand, the power plant has to be adjusted to electrical curve, which was presented in Fig. 7.13.

The result of these temporal variations hinders the possibility to minimize the energy consumption. This section deals with this subject to demonstrate how tricky might be to design by averaged values instead of analyzing adjustment in real time.

C. Cuviella-Suárez et al., *Water and Energy Use in Sanitary-ware Manufacturing*, Green Energy and Technology, https://doi.org/10.1007/978-3-030-72491-7_8

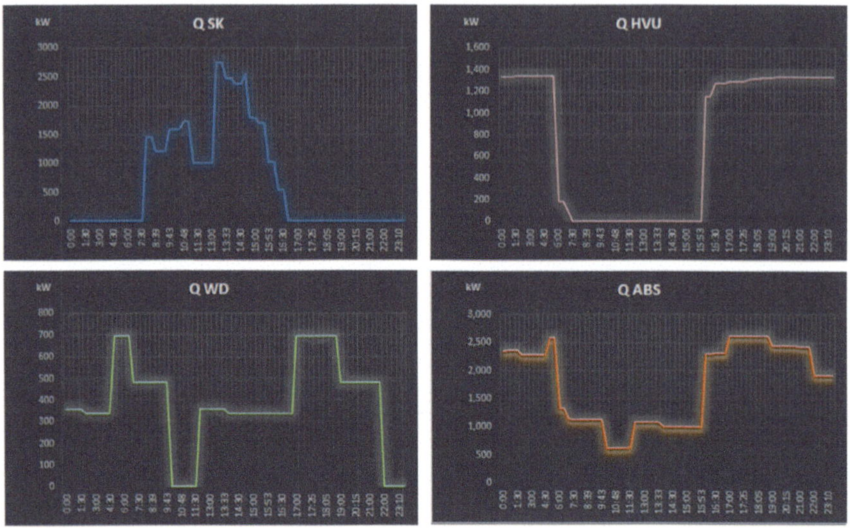

Fig. 8.1 Thermal demand of the variable thermal consumers along the daily cycle (Shuttle kiln, heating and ventilation unit and ware dryer)

8.2 Cycle Analysis

Following the stated models, the thermal powers of each cycle are computed over a period of 24 h.

Though some consumers operate on a basis of variability, some other can be considered as in a steady condition. Such is the case for the tunnel kiln, mold dryer, MED distiller or absorption machine. The procedure for the analysis has been to break down the period cycle of 24 h into smaller periods matching with the different individual cycles. On this regard, the curve of the shuttle kiln defines the time partition due to its greater variability. Cycle ranges for the stages of the different machines are; 2 or 3 h for the dryer, 30 min for the HVU. Regarding the gas turbine, it has been stated 1 h as analysis period.

After these considerations and by means of each mathematical model, it has been calculated the thermal requirement of natural gas and the recoverable amount of heat for every single period of time to complete the whole cycle of 24 h. It has been carried out primarily, for a conventional factory where no saving measures have been carried out. The set of equations for this purposes is:

$$Q_{NG} = \sum_i Q_{NGi} \tag{8.1}$$

$$Q_{RH} = Q_{RHTK} + Q_{RHSK} \tag{8.2}$$

$$m_{CO_2} = \sum_i m_{CO_2 i} \tag{8.3}$$

$$m_{H_2O} = \sum_i m_{H_2Oi} \tag{8.4}$$

$$m_{air+N_2} = \sum_i m_{air+N_2 i} \tag{8.5}$$

$$Q_{ABS} = \frac{\left(m_{H_2O_{MD}} + m_{H_2O_{HVU}} + m_{H_2O_{WD}}\right) \cdot (h_v - h_l)}{COP_{ABS}} \tag{8.6}$$

where:

i: each thermal consumer

NG: Natural gas required to fuel the equipment

RH: Recovered heat from tunnel and shuttle kilns

Q: Thermal power required for the *i*th consumer at every stage of the cycle

m: mass of CO_2, H_2O and residual gases (air + N_2) within the exhausting of each consumer

h_v: Enthalpy of vapor (kJ/kg)

h_l: Enthalpy of condensate (kJ/kg)

COP_{ABS}: Coefficient of performance of the absorption machine (kW_{cold}/kW_{heat}).

By the energy equations are calculated the fuel requirement so as the recoverable amount of heat to be reutilized for equipment downwards in a temperature scale. On the other hand, the exhausting masses allow to compute CO_2 emissions and recoverable amount of water via absorption chiller.

Resultant values are summarized in Fig. 8.2.

The analysis of this figure suggests that recovered heat from kilns is enough to fulfill the complete thermal requirements of the rest of the consumers. This fact might induce to add any action to recover water with the excess of his heat. So, a MED distiller and an absorption machine are implemented to achieve it. The result of this is that recovered heat does not reach the new demand and the natural gas consumption increases. By recovering the heat from kilns, thermal efficiency would improve a 50% in a conventional factory and 35% in that with water-saving measures. However, when the analysis goes deeper into temporal variations, those averaged values that seemed to supply other consumers are revealed to fail during some hours. On the contrary, there are some other periods when there is a lack of recovered heat. These facts are clearly visible in Figs. 8.3 and 8.4.

Figure 8.3 shows a clear excess of recovered heat during day time and a slight lack during night because of the strong demand to dry molds in the casting hall.

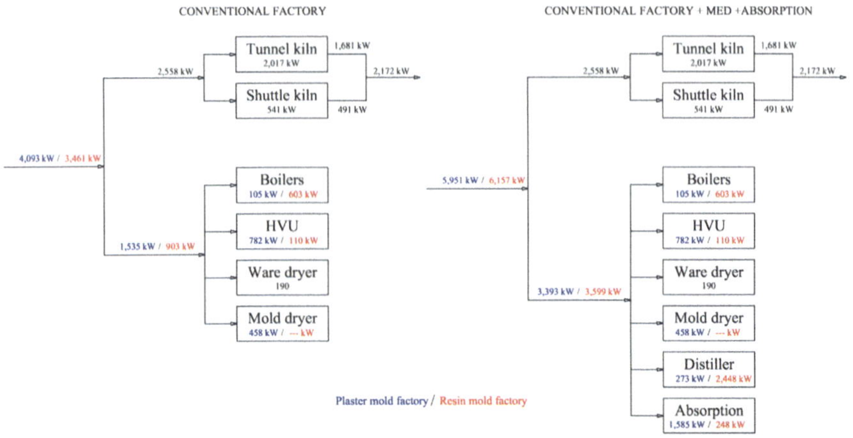

Fig. 8.2 Schemes of a simple conventional factory and extension to recover water by distillation and absorption

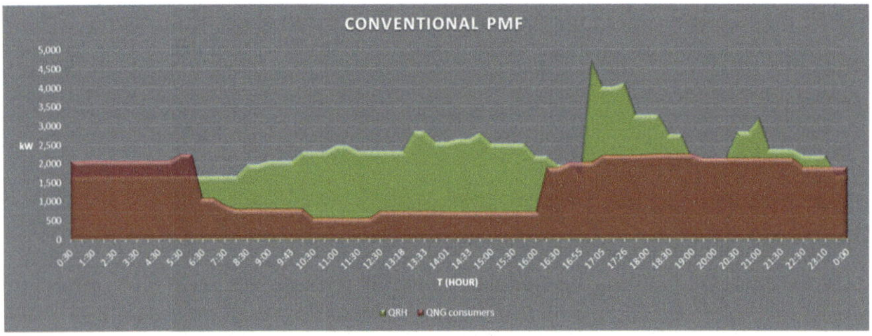

Fig. 8.3 Conventional plaster mold factory

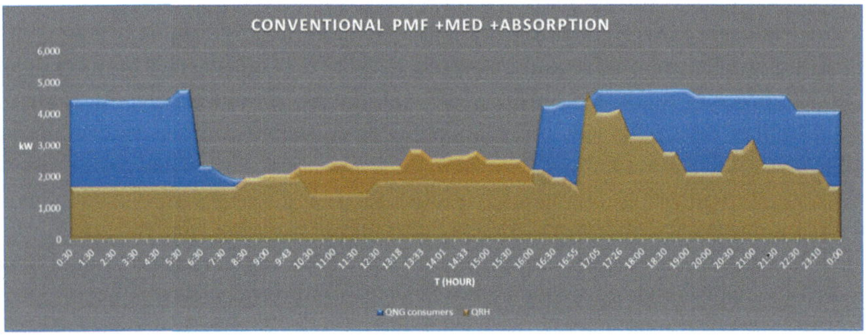

Fig. 8.4 Conventional plaster mold factory with the addition of a MED distiller and an absorption machine

On the other hand, Fig. 8.4 agrees with the averaged values to demonstrate that recovered heat fails to fulfill total thermal supply to the rest of consumers. Only 60% can be supplied.

At this point, it is suggested to implement a power plant as designed in 7.3.8. which results in an excess of recovered heat from the gas turbine to supply the conventional consumers.

Gas turbine contribution to energy and flue gases has to be added to equations system as:

$$Q_{NG} = \sum_i Q_{NGi} \tag{8.7}$$

$$Q_{RH} = Q_{RHGT} + Q_{RHTK} + Q_{RHSK} \tag{8.8}$$

$$m_{CO_2} = \sum_i m_{CO_2 i} \tag{8.9}$$

$$m_{H_2O} = \sum_i m_{H_2Oi} \tag{8.10}$$

$$m_{air+N_2} = \sum_i m_{air+N_2 i} \tag{8.11}$$

$$Q_{ABS} = \frac{\left(m_{H_2O_{MD}} + m_{H_2O_{HVU}} + m_{H_2O_{WD}}\right) \cdot (h_v - h_l)}{COP_{ABS}} \tag{8.12}$$

Figure 8.5 summarizes the energy resultant values.

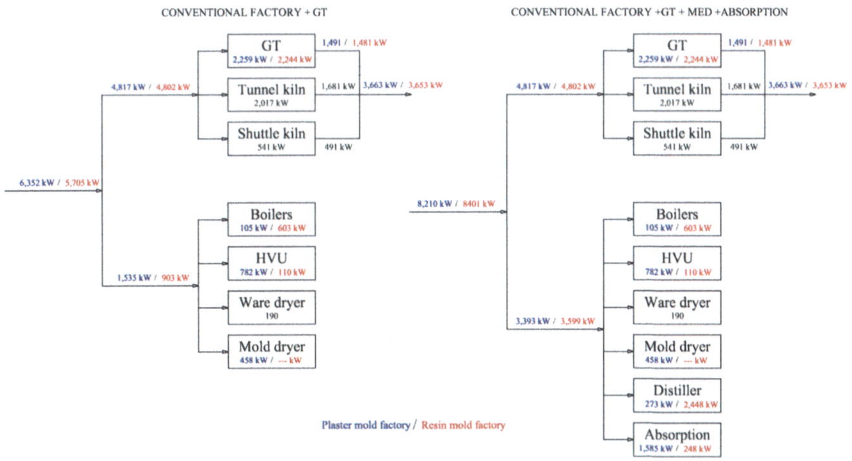

Fig. 8.5 Schemes of a simple conventional factory and extension to recover water by distillation and absorption with the addition of a cogeneration plant

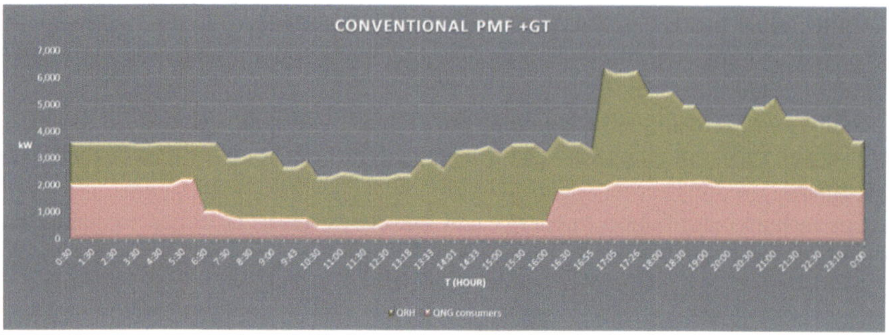

Fig. 8.6 Conventional plaster mold factory with the addition of a cogeneration plant

In Fig. 8.6, it is clearly stated the great excess of wasted heat for the first configuration.

Under such circumstances, it is more than advisable to add the water-saving measures. Hence, the curves of demand and recovering are better matched than Fig. 8.7. This is going to be the start point to optimize the factory. In Chap. 9, improvements go further by applying the heat recovery and recalculate the thermal balances.

These graphics have been computed for a plaster mold factory. By means of the same procedure, they might be produced for the resin mold factory, Figs. 8.8, 8.9, 8.10 and 8.11 show them. Recoverable heat is essentially the same. The thermal demand is slightly different as seen in Figs. 8.2, 8.3, 8.4 and 8.5.

In order to summarize the recovery potential, Table 8.1 shows the same trend of heat recovery than previous figures.

From the table, it can be concluded that natural gas consumption has been increased with respect to conventional factory more than 100% from the perspective of the site. However, when deeper analysis considering the fuel required at the power station, and the losses through power transmission system, the balance is positive.

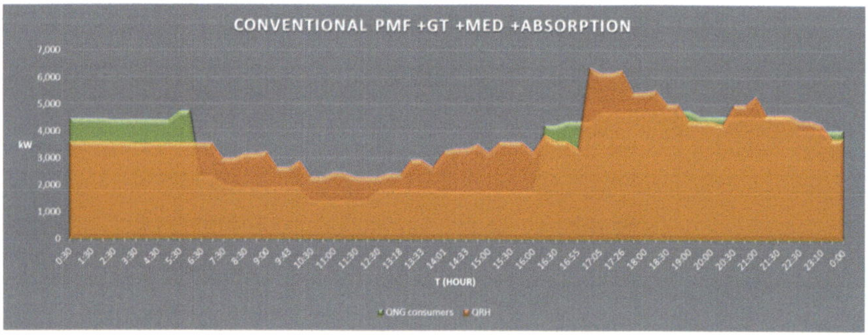

Fig. 8.7 Conventional plaster mold factory improved with the addition of the cogeneration plant, MED distiller and absorption machine

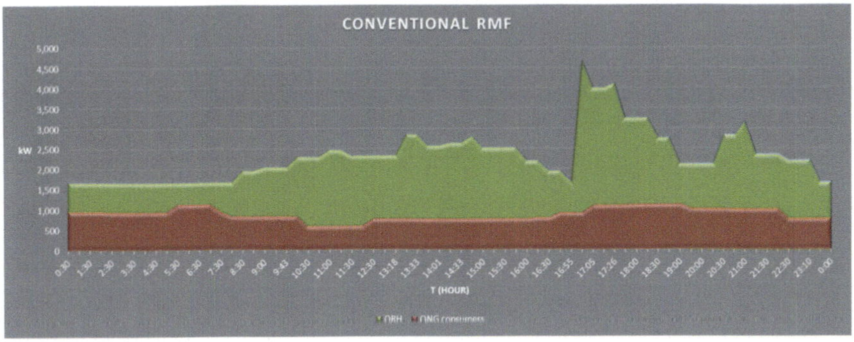

Fig. 8.8 Conventional resin mold factory

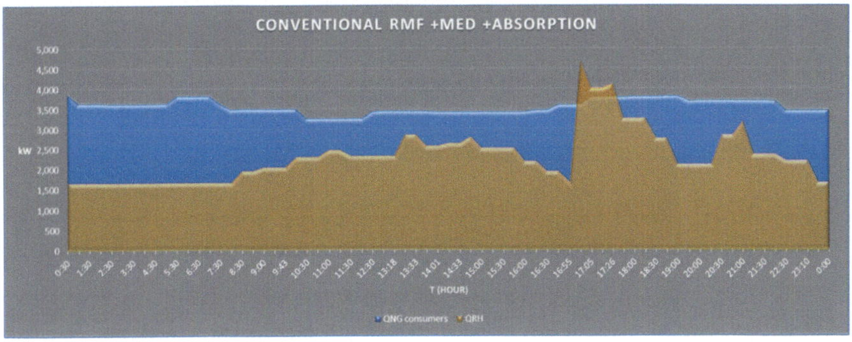

Fig. 8.9 Conventional resin mold factory with the addition of a MED distiller and an absorption machine

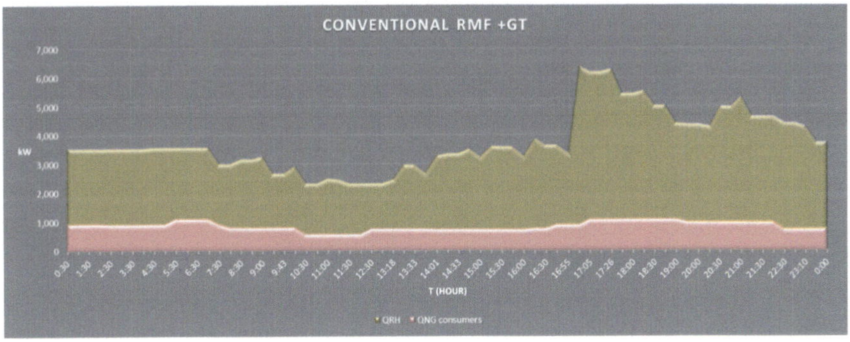

Fig. 8.10 Conventional resin mold factory with the addition of a cogeneration plant

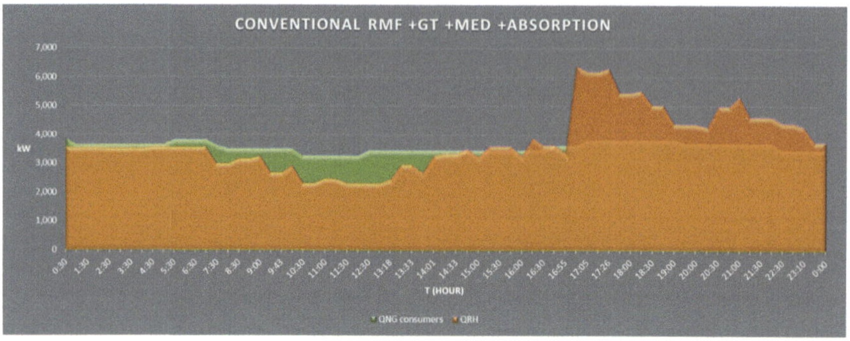

Fig. 8.11 Conventional resin mold factory with the addition of the cogeneration plant, MED distiller and absorption machine

Table 8.1 Averaged thermal power of the consumers besides the kilns versus recovered heat

Potential of heat recovery (kW)		
	PMF	RMF
Conventional factory		
Q_{NG}	4,093	3,460
Q_{NG} consumers	1,534	902
Q_{RH}	2,172	2,172
Conventional factory +MED +ABS		
Q_{NG}	5,951	6,156
Q_{NG} consumers	3,393	3,597
Q_{RH}	2,172	2,172
Conventional factory +GT		
Q_{NG}	6,352	5,704
Q_{NG} consumers	1,534	902
Q_{RH}	3,663	3,653
Conventional factory +GT + MED + ABS		
Q_{NG}	8,210	8,399
Q_{NG} consumers	3,393	3,597
Q_{RH}	3,663	3,653

Besides, when the electricity is generated on site, there is the benefit of heat recovery, what in this case is very useful to recover also water. The recoverable heat has been increased by 68% due to the addition of the GT and consumers demand has risen up to 21% for the PMF and 200% for the RMF because of MED distiller and absorption cycle. This fact allows to take profit of the extra residual heat which, otherwise would be waste at the power station. It is important to remark that this exercise has been carried out assuming that both, distiller and absorption are fueled by natural gas. The aim of the study is to allow recovered heat instead, as described in Chap. 9.

8.3 Water Condensation Potential and Co$_2$ Emissions

The same line of reasoning is applied to water and CO_2 contents within the exhausting to find out how much water may be condensed and the impact of CO_2 emissions when new consumers are added to the conventional factory. Table 8.2 and Fig. 8.12 show, respectively, the average and variation of these contents.

Corresponding graphs of temporal variations of CO_2 and H_2O emissions are presented in Figs. 8.12 and 8.13.

With regard to GT configuration, it has to be remarked the valley in the central part of the day due to photovoltaic contribution to power generation. This fact is clearly visible in the CO_2 figure. H_2O hides this effect because the great contribution of the HVU in the plaster factory overlaps onto the GT shape showing very clearly the shape of the HVU curve with an abrupt step between drying and casting periods. Resin mold factory HVU curve is steadier because there is a unique HVU stage.

It is important to mention the fact that this study only has considered condensation from evaporated water through mold and ware dryers so as heating and ventilation units. It has not been considered all the vapor that might be condensed from combustion processes in the gas turbine and kilns. This is an important amount that might be recovered as far the residual heat allows for it.

Table 8.2 Averaged composition of flue gases depending on the configuration of facilities

Flue gas composition (kg/s)		
	PMF	RMF
Conventional factory		
m_{CO_2}	0.2344	0.1981
m_{H_2O}	0.5899	0.2523
m_{air}	17.9399	9.5594
Conventional factory +MED +ABS		
m_{CO_2}	0.3408	0.3524
m_{H_2O}	0.6769	0.3785
m_{air}	18.0613	10.4269
Conventional factory +GT		
m_{CO_2}	0.3638	0.3266
m_{H_2O}	0.6958	0.3574
m_{air}	21.3831	12.9791
Conventional factory +GT + MED + ABS		
m_{CO_2}	0.4702	0.4809
m_{H_2O}	0.7828	0.4837
m_{air}	21.5045	13.8466

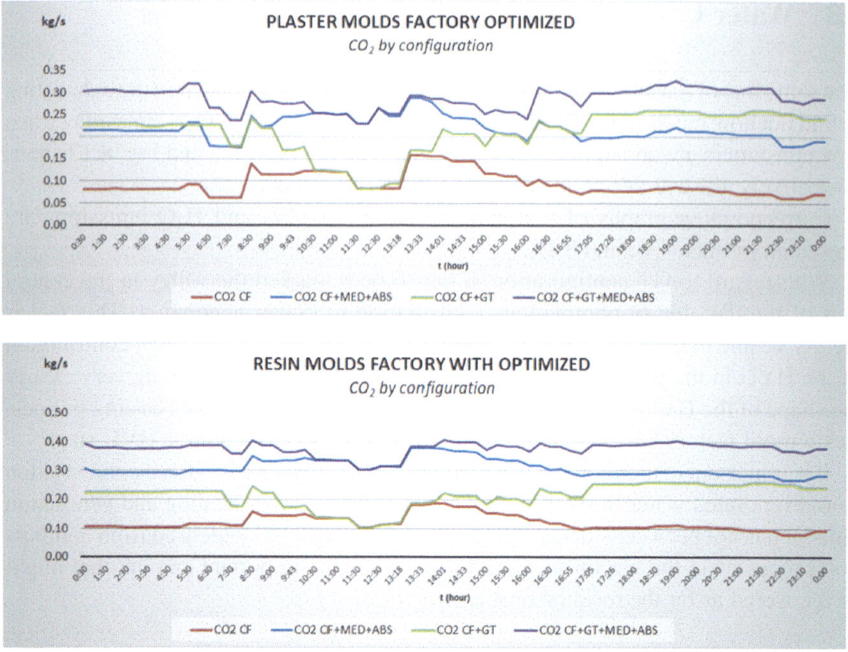

Fig. 8.12 Variation of CO_2 emissions for a daily cycle

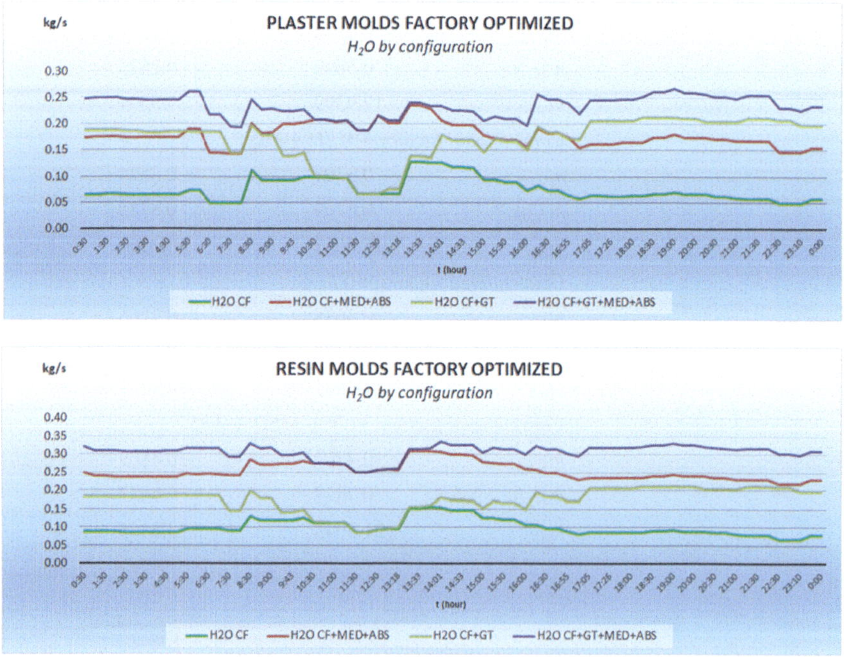

Fig. 8.13 Variation of H_2O emissions for a daily cycle

8.4 Recovery Strategy Conclusions

In this section, actual potential of recovery has been demonstrated, for energy as well as for water. Nevertheless, there have been revealed some restrictions regarding the total efficiency improvement deriving from differences in demand and supplying curves.

It has been stated that over one third of the natural gas consumption could be potentially reduced by recovering heat from the kilns toward other heat consumers for a simple conventional factory. It has been demonstrated that it can be increased up to almost a half by just heat recovery. In [1], it is explored the potential of waste heat recovery in the EU ETS glass industry according to the latest BAT for the ceramic industry.

This study aims to maximum efficiency so it has not limited the savings to heat recoverig techniques [2]. Hence, there has been proposed an innovative way to further improve electrical and water efficiencies. As a mean to reduce environmental impact of the industry, it has been implemented a co-generation as a novel technique to improve the performance of this sector. There are other ceramic industrial specialties, which have already achieved the improvement by this system [3]. On the other hand, it is worth considering the economic impact of thermal recovering [4], but this matter will be treated in Chap. 11.

References

1. Karellas S, Giannakopoulos D, Hatzilau C-S, Dolianitis I, Skarpetis G, Zitounis T (2018) The potential of WHR/batch and cullet preheating for energy efficiency in the EU ETS glass industry and the related energy incentives. Energy Effic 11(5):1161–1175. https://doi.org/10.1007/s12053-017-9587-3
2. Broberg Viklund S (2015) Energy efficiency through industrial excess heat recovery—policy impacts. Energy Effic 8(1):19–35. https://doi.org/10.1007/s12053-014-9277-3
3. Caglayan H, Şöhret Y, Caliskan H (2018) Thermo-ecologic evaluation of a spray dryer for ceramic industry. Energy Proc 144:164–169. https://doi.org/10.1016/j.egypro.2018.06.022
4. Colmenar-Santos A, Rosales-Asensio E, Borge-Diez D, Collado-Fernández E (2016) Evaluation of the cost of using power plant reject heat in low-temperature district heating and cooling networks. Appl Energy 162:892–907. https://doi.org/10.1016/j.apenergy.2015.10.161

Chapter 9
Optimized Factory Versus Conventional Factory

9.1 Optimized Configuration

In Sect. 5.5, Fig. 5.8 was stated the consumptions for a conventional factory, which are shown (Fig. 9.1).

In Sect. 7.3.8, a cogeneration plant was proposed in order to provide the electricity out of photovoltaic supply. An excess of residual heat may result, which can be profited to feed the proposed water-saving measures. Figure 9.2 shows the proposed configuration of the optimized factory where all the exclusive saving measures have been applied after reducing consumptions via general saving measures.

The averaged consumption values for the period of 24 h. are shown.

This optimized factory allows for all described measures so far. However, the inclusion of water-saving measures implies an increase of energy consumption that might be assumable depending on different factors as; energy cost, water cost, environmental responsibility, water source availability, etc.

9.2 Results and Discussion

When the model simulators are run through a complete cycle of 24 h, the variability of consumption can be plotted, which most relevant results are shown in Figs. 9.5, 9.6 and 9.7.

The residual heat from the kilns has been reutilized to preheating fresh air to feed combustion processes inside their chamber or by forcing draught air within the tunnel kiln upwards to contribute this preheating of the running load inside. This fact has produced a decrease in residual heat to be recovered. On the other hand, the inclusion of the gas turbine has produced an increase of residual heat that over pass the production requirement. The turbine has been designed to fulfill the electrical requirement further the photovoltaic scope in order to allow the factory to be disconnected from the electrical network (Figs. 9.3 and 9.4).

C. Cuviella-Suárez et al., *Water and Energy Use in Sanitary-ware Manufacturing*, Green Energy and Technology, https://doi.org/10.1007/978-3-030-72491-7_9

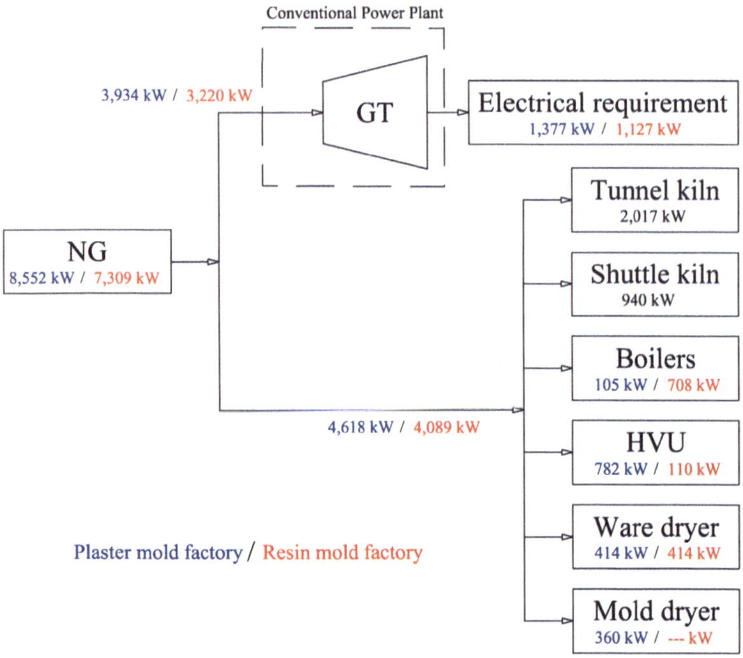

Fig. 9.1 Scheme of thermal consumptions distribution in a conventional factory (*Source* Data compiled by the authors)

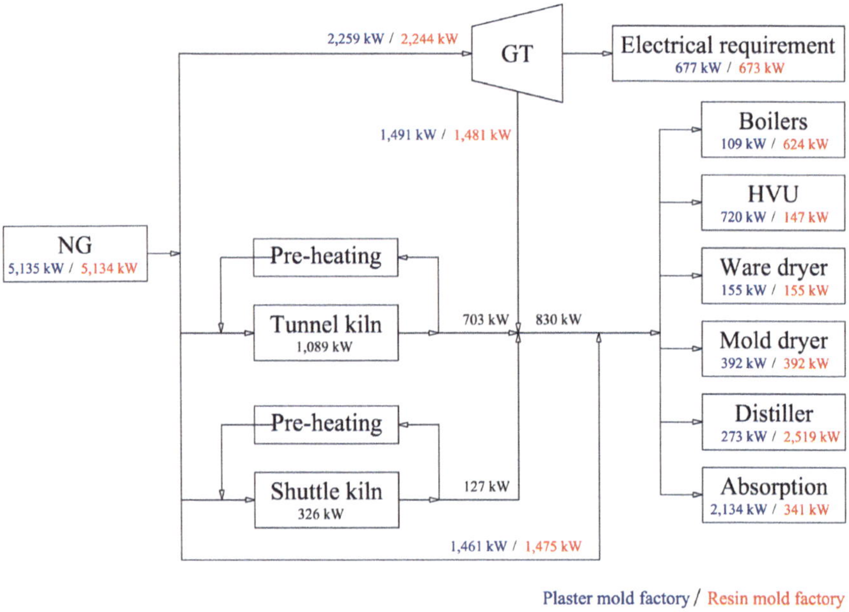

Fig. 9.2 Scheme of thermal consumptions distribution in an optimized factory

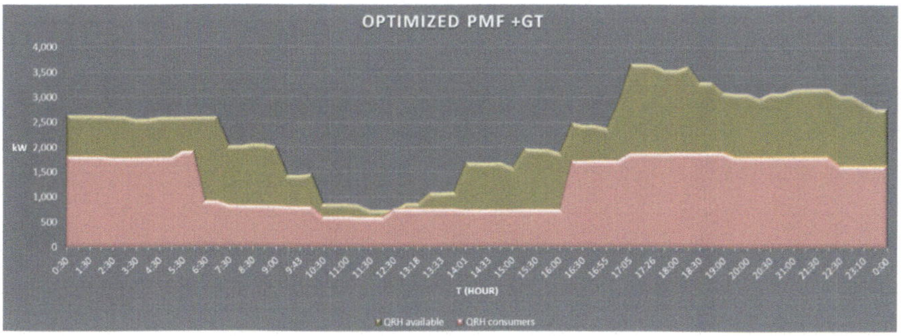

Fig. 9.3 Excess of residual heat when the GT is applied to the optimized plaster mold factory (*Source* Data compiled by the authors)

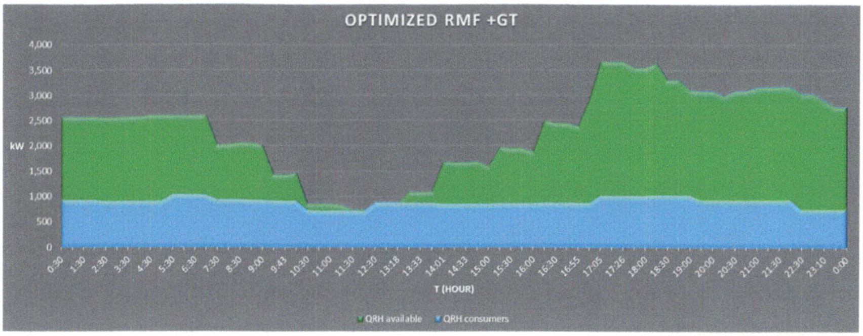

Fig. 9.4 Excess of residual heat when the GT is applied to the optimized plaster mold factory (*Source* Data compiled by the authors)

The excess of residual heat is clear in both figures. Since the main aim of this book is the total optimization of resources, it is mandatory to find the way to reuse this excess of energy. It has been applied to recover water by means of thermal processes in the search of the recovery optimization (Figs. 9.5 and 9.6).

It appears an inverse situation regarding the availability of residues where a clear lack of residual heat appears from the great demand of heat to either distillate or condense residual water. The optimal situation from the environmental point of view would be to adjust the amount of water to be recovered defined by the exact excess of heat at a very moment along the daily cycle of thermal demand. The variation of emissions is shown in Fig. 9.7. A PMF produces more CO_2 than the RMF because of greater thermal demand to dry molds during mold making process and also for mold drying after casting. H_2O emission is greater for the same reason.

From an economy point of view, cost of energy and water should be included within the optimization calculations in order to minimize global cost. This effect will be analyzed in Chap. 11.

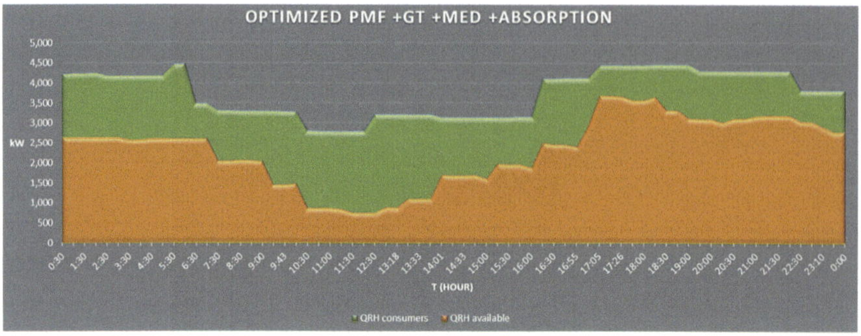

Fig. 9.5 Availability of recovered heat to be reused through thermal consumers in an optimized plaster mold factory (*Source* Data compiled by the authors)

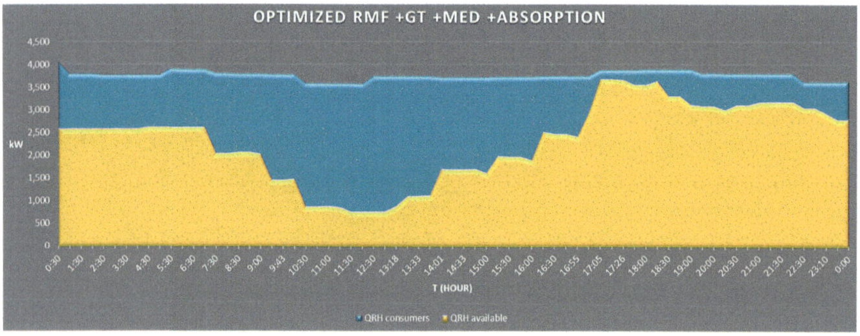

Fig. 9.6 Availability of recovered heat to be reused through thermal consumers in an optimized resin mold factory (*Source* Data compiled by the authors)

Table 9.1 shows the single results regarding heat recovery for both, heat providers and consumers.

These values have been aggregated following the optimized scheme of Fig. 9.2 to compute the averaged performance of the optimized factory to be confronted with conventional configuration and corresponding consumptions. A fare comparison must consider the primary energy to produce electricity at the power plant of the electrical network. It has been added to the comparison table as well as corresponding CO_2 emissions (Table 9.2).

This summary of savings shows that electrical reduction has been total from the electrical network point of view as a combination of actual reduction of consumptions and an on-site power production. Water measures have not been included within yet to be able to show the excess of recoverable heat that is produced. They are computed from Table 9.1. Leftovers of residual heat equal 946 and 1,385 kW, respectively, for a PMF and RMF. In the search of the optimization, Table 9.3 shows the results when considering MED and absorption. In this case, the lack of residual heat results to be 1,461 and 1,475, respectively, for PMF and RMF.

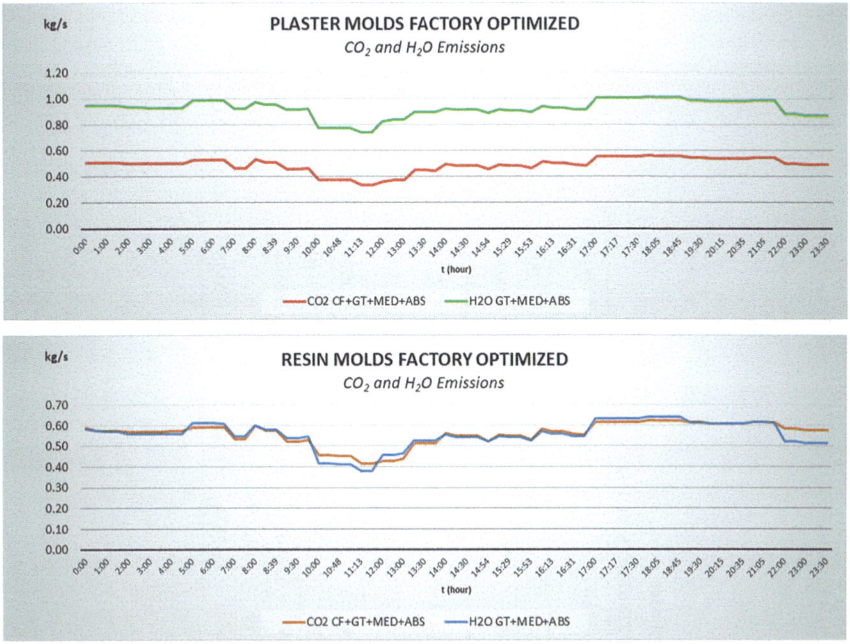

Fig. 9.7 CO_2 and H_2O emissions for optimized factories

Comparing both configurations each other, it can be seen how far the efficiency can be improved. Before applying water measures, the cogeneration plant improves the thermal efficiency about 10% for both type of factory with respect to the conventional one. CO_2 emissions are cut down further because of the general reduction measures. These reductions reach up to almost 40%. Nevertheless, when trying to optimize heat recovery, there is an increase in thermal consumption that is almost negligible in a PMF (0.92% of thermal requirement increase). However, it reaches 25% in the case of the PMF.

The difference between both configurations suggests that a commitment between them might be the optimal solution, which would take just the excess of heat to recover water. Under such circumstances, water recovery would not be complete, but no extra fuel consumption would be required.

The exercise of reduction has been a success in order to optimize consumption and environmental impact. Nevertheless, there is still an important amount of emissions that should be cut down. At the current state of the art, it is very difficult to eliminate completely the CO_2 emissions, since there is not a proven technology to fire glazed ceramics under industrial requirements. There is a lack of knowledge regarding this process in order to allow it to be fueled by green energy, such as renewable electricity. Alternatives to natural gas combustion might be:

- H_2 combustion.
- Electrical kins supplied by renewable energies

Table 9.1 Distribution of thermal consumptions and heat recovery for the optimized factories

	Optimized PMF (Heat recovery)						OPTIMIZED RMF (Heat recovery)					
RH provider	Q_{NG} *(kW)*		Q_{RH}	CO_2 *(kg/s)*	H_2O *(kg/s)*	*Air+*N_2 *(kg/s)*	Q_{NG} *(kW)*		Q_{NG} *(kW)*	CO_2 *(kg/s)*	H_2O *(kg/s)*	*Air+*N_2 *(kg/s)*
GT	2,259		1,491	0.1294	0.1059	3.4432	2,244		1,481	0.1285	0.1052	3.4197
TK	1,089		703	0.0624	0.0511	3.6756	1,089		703	0.0624	0.0511	3.6756
SK	326		127	0.0185	0.0151	0.4223	326		127	0.0185	0.0151	0.4223
RH consumer	Q_{RH} *(kW)*	m_{RH} *(kg/s)*	$Q_{exhausting}$ *(kW)*	CO_2 *(kg/s)*	H_2O *(kg/s)*	*Air+*N_2 *(kg/s)*	Q_{RH} *(kW)*	m_{RH} *(kg/s)*	$Q_{exhausting}$ *(kW)*	CO_2 *(kg/s)*	H_2O *(kg/s)*	*Air+*N_2 *(kg/s)*
MD	392	0.1458	136	0.0168	0.1342	2.6940						
BO	109	0.4482	47	0.0107	0.0087	0.4288	624	2.5739	270	0.0614	0.0502	2.4623
HVU	720	2.3612	1,772	0.0901	0.4214	16.9917	147	3.2964	207	0.0110	0.0494	3.2362
WD	155	0.6771	283	0.0162	0.0684	1.7625	155	0.6771	271	0.0162	0.0684	1.5626
MED	273	1.0942	48	0.0261	0.0214	1.0467	2,519	10.1099	445	0.2411	0.1973	9.6715
ABS	2,134	7.8481	924	0.1223	0.1000	0.0339	341	1.2630	148	0.0196	0.0160	0.0054

Table 9.2 Comparison of conventional and optimized factories before implementing MED and absorption machine. (*Source* Data compiled by the authors)

	Plaster mold factory				Resin mold factory			
	Electrical	Thermal	Water	CO_2	Electrical	Thermal	Water	CO_2
	kWh/year	kWh/year	m^3/year	t/year	kWh/year	kWh/year	m^3/year	t/year
Conventional factory consumption	7,669,258	28,110,353	32,930	10,317	6,277,498	23,481,912	242,496	8,542
Equivalent thermal fuel to electricity		21,912,165		4,519		17,935,707		3,699
Global conventional factory	7,669,258	50,022,518	32,930	14,837	6,277,498	41,417,619	242,496	12,242
Actions (savings and increases)								
Electrical general measures	1,921,782			1,132	1,502,813			886
Thermal general measures		4,470,122		922		3,820,606		788
Induced draught and pre-heating for tunnel kiln		6,680,739		1,378		6,680,739		1,378
Pre-heating for shuttle kiln		1,931,509		398		1,931,509		398
Photovoltaic plant	2,104,488			1,240	2,039,380			1,202
Cogeneration plant	3,642,987	−20,714,034		−2,126	2,735,304	−20,708,761		−2,659
Heat recovery		12,923,385		2,665		12,866,832		2,654
Optimized factory consumption	0	44,730,796	32,930	9,226	0	36,826,694	242,496	7,596
Efficiency improvement	100%	10.58%	0.00%	37.82%	100%	11.08%	0.00%	37.95%

Table 9.3 Comparison of conventional and optimized factories when water-saving measures have been applied

	Plaster mold factory				Resin mold factory			
	Electrical	Thermal	Water	CO_2	Electrical	Thermal	Water	CO_2
	kWh/year	kWh/year	m^3/year	t/year	kWh/year	kWh/year	m^3/year	t/year
Conventional factory consumption	7,669,258	28,110,353	32,930	10,317	6,277,498	23,481,912	242,496	8,542
Equivalent thermal fuel to electricity		21,912,165		4,519		17,935,707		3,699
Global conventional factory	7,669,258	50,022,518	32,930	14,837	6,277,498	41,417,619	242,496	12,242
Actions (savings and increases)								
Electrical general measures	1,921,782			1,132	1,502,813			886
Thermal general measures		4,470,122		922		3,820,606		788
Induced draught and pre-heating for tunnel kiln		6,680,739		1,378		6,680,739		1,378
Pre-heating for shuttle kiln		1,931,509		398		1,931,509		398
Photovoltaic plant	2,104,488			1,240	2,039,380			1,202
Cogeneration plant	3,766,843	−20,714,034		−2,053	3,650,306	−20,708,761		−2,120
Heat recovery		12,923,385		2,665		12,866,832		2,654
Absorption machine	−27,840	−4,277,379	4,412	−899	−27,840	−1,348,703	1,391	−295
MED distillation	−96,016	−1,475,110	12,842	−361	−887,161	−13,629,616	118,661	−3,334
Optimized factory consumption	0	50,483,285	15,675	10,412	0	51,805,013	122,444	10,685
Efficiency improvement	100%	−0.92%	52.40%	29.82%	100%	−25.08%	49.51%	12.72%

- CO_2 capture and treatment before releasing it to the atmosphere.
- …

This must be the challenge for this industry to be developed in short term to become really sustainable.

Chapter 10
Exergoeconomic Analysis

10.1 Introduction to the Analysis Approach

The approach of this chapter is based on [1] as a baseline. The cost of thermal energy represents one of the largest items of production cost of ceramic manufacturing and the CO_2 emissions resulting from this industrial activity are very significant. Only the tile sector, in Brazil, represents 1% of the GDP (gross domestic product), 1.9% of the total consumed energy and 5.8% of the industrial consumed energy [2]. Any improvement in the efficiency of thermal processes will result in a substantial reduction of costs of manufacturing and CO_2 emissions. On the other hand, water is rising as a scarce resource though the impact on the cost is not so high by now. There is a direct correlation between global energy development and climate-induced water scarcity [3]. Coupled water-energy policies must enforce the application of measures addressed to the reduction of these resources. Given the growing importance of these issues, the environment and competitiveness, during the last years, there has been a significant amount of information on the consumption of thermal energy in the sector of ceramics [4]. In this scenario, it is essential to obtain information that allows establishing correlations between the characteristics of the industry (manufacturing process, characteristics of the facilities, operating conditions, main raw materials, etc.) and the thermal consumption in each stage of the process and the total emissions of CO_2. There is much research regarding industrial processes from the exergy point of view of exergy analysis. Most of this research aims to individual single industrial processes. However, there is little research concerning a complete industrial sector from that perspective. Vuckovic et al. analyze an industrial chemical process in [5]. Exergoeconomic analysis is a powerful tool to find opportunities of energy reduction and therefore, cost savings. It integrates energy, environment, sustainable development and industrial ecology [6]. Application of heat transfer analysis to the overall thermal process and the thermal links between sub-processes for manufacturing sanitary-ware are the most practical ways to figure out the most suitable reduction of energy consumption. Kita et al. have studied it in [7], in green ceramics processing. The authors have found much research regarding

C. Cuviella-Suárez et al., *Water and Energy Use in Sanitary-ware Manufacturing*, Green Energy and Technology, https://doi.org/10.1007/978-3-030-72491-7_10

ceramic industry in general and tile manufacturing in particular, but little articles about sanitary-ware industry from the perspective of energy and water consumption. In [8], it can be found an exergetic analysis of a roller kiln for firing tiles. It states an efficiency of 15% and 83% of the exergy is destroyed. Over 61% of the waste heat can be recovered in order to increase performance. In the same way that there, the present research takes the analyses of the kilns to identify the most critical parameters affecting kiln energy efficiency and proposes ways of improving kiln energy performance. An immediate way to save energy in the kilns is the pre-heating of the combustion air. When the combustion air is heated up externally to the kiln, this amount of energy is saved from the kiln consumption. In this sense, both natural gas and CO_2 emissions result reduced. Caglayan et al. state that 4.7% of energy is reduced by heating combustion air up to 148 °C and 8–10% in the range of 200–250 °C [9].

Proper thermal insulation of the kilns reduces the thermal losses through enclosure. This is another way to reduce consumption. Again, exergoeconomic analysis is able to find out how achievable is to reach the optimum from an exergoeconomic point of view [10].

In the framework of CO_2 reduction in Europe, there is some research regarding alternatives for ceramic tiles manufacturing [11] where up to 17 technological alternatives and their combinations propose 25 technological scenarios to attempt this subject.

Regarding drying process, it has been found some research within the tiles sector as well [12]. This technology is not suitable for sanitary-ware since there has not been developed yet a proper process of dry molding for the sanitary ware manufacturing. Some research is being carried out at present [13], but still not industrially developed. In [14], Caglayan et al. analyze spray drying from an exergetic point of view since it can represent up to 43% of the total thermal consumed energy.

Sanitary-ware as any ceramic process implies great amounts of energy and water along the manufacturing line. The document on best available technologies for the ceramic industry [15] states that energy to produce sanitary ware has been improved from 26.56 MJ/kg in 1,980 to 21.87 in 2,003. Up to 116,640 TJ/year are currently consumed to produce sanitary ware worldwide [16]. A modern tunnel kiln for firing sanitary-ware consumes 4.2 up to 6.7 MJ/kg, which makes about 20–30% of the total thermal consumption. A shuttle kiln takes 8.3–10.4 MJ/kg. Fortunately, this process only is carried out for rejected ware through the quality control. On the other hand, drying process requires 0.3–1.4 MJ/kg, which represents 1.3–6.4%.

There are some attempts to study water saving policies in industrial areas as Bi et al. done in [17]. The novelty of this research lays on the specificity of the sanitary-ware industrial sector from the exergoeconomic perspective to reinforce the energy–water nexus by engaging a combination of engineering techniques to reuse wastes of heat and water into a linked reduction of both resources. The water–energy nexus to improve the consumption of the single factory does not require the support of shared networks. It has not been approached from this perspective before. This study can be considered as a practical tool to identify the chances of energy and water reduction, quantify the avoidable exergy destruction and recover this exergy in order to reduce the global energy requirement so as the water need by re-utilization via distillation [18]. Any action aiming reduction of resources is advisable to keep a sustainable industry and minimizing the environmental impact [15]. As stated in [19], the IR_3 concept (infinite reusable, recyclable and renewable industrial ecosystem) proposes an infinitely reusable, recyclable, and renewable industrial ecosystem with no waste to download.

Yet there is much research regarding energy and water saving measures and recommending them, sometimes there are restrictive factor to follow the recommendations or take on the investments. Cantareli et al. have analyzed them in [20]. As a result, 19 factors were considered as restrictive among 28 identified factors.

The solving procedure starts from the conventional factory with an exergoeconomic analysis to quantify the exergy that can be recovered and to identify the possible ways to carry it out. Then the configuration of the factory is optimized by means of thermal engineering methods to recover residual heat from some sub-processes to be profited in other ones, so as to distillate water. Finally, both configurations are compared in order to show the global savings and the results are compared with the existing data regarding this industry.

In order to allow for the calculation, thermodynamic models have been developed for every component of the system. In this sense, the computational modeling allows the researcher to make predictions about the results that could be tested later on in a second stage of research [21]. In the first stage, the results have to be coherent with existing data. In the area of hybrid simulation [22], the research provides a tool to better evaluate the relationship between actual existing data in the sanitary-ware manufacturing and the parameters that allow to optimize the performance of the efficiency of resources.

Solving procedure flow chart is shown in Fig. 10.1.

In a CF (conventional factory), all the thermal processes are led by NG (Natural Gas) supply, electricity from the network, through a conventional power plant, and water from the network, as stated in Sect. 9.1, Fig. 9.1.

On the other hand, all residual heat is delivered to the atmosphere carrying certain degree of exergy, which might be profited for other purposes when properly analyzed [23].

Both factory conceptions are compared by advanced exergy analysis in order to identify and quantify the exergoeconomic reduction when the CF is optimized into OF (Optimized factory). Energy and exergy analysis of equipment has been carried out individually in order to characterize every one of them from the thermodynamic point of view establishing the complete map of flows among them for the two configurations. The exergetic analysis has been computed for standard conditions of efficiency so as for the best possible conditions in order to establish the unavoidable destroyed exergy regarding both factory configurations. After this, exergoeconomic analysis has been achieved to demonstrate that both primary energy and water consumption can be cut down at once deriving an important environmental and economic improvement, which implies a fundamental impact of sustainability for this industrial sector.

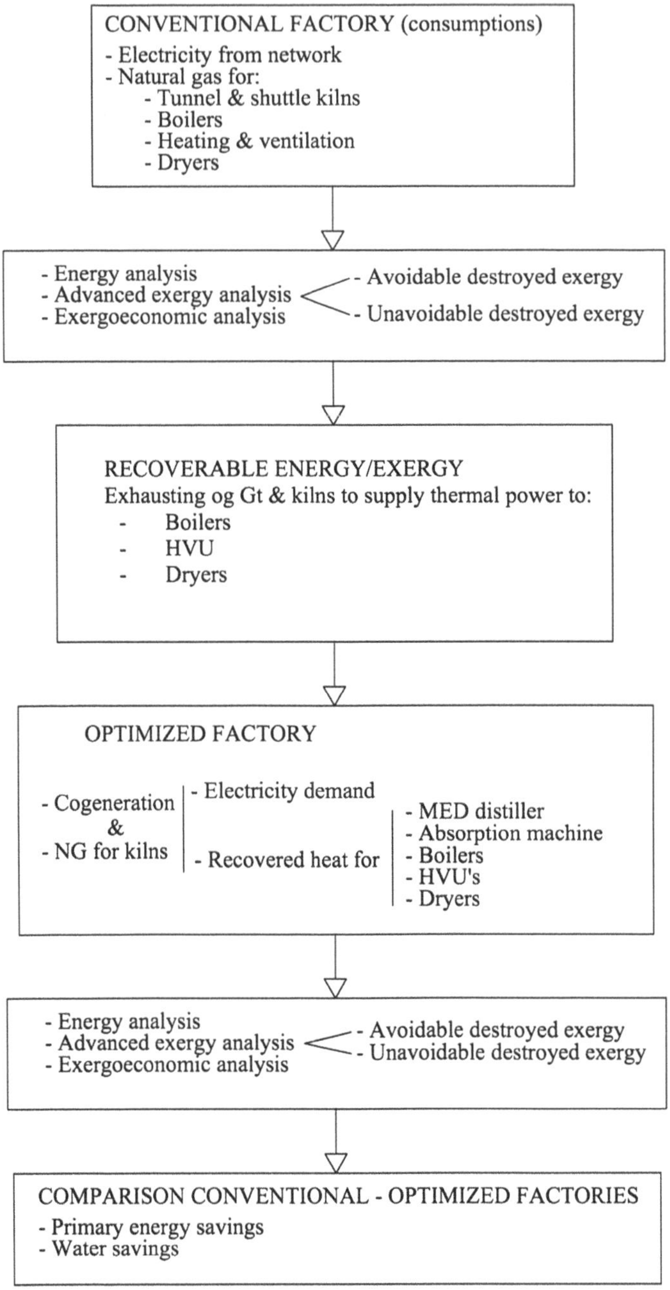

Fig. 10.1 Flowchart of solving procedure

10.2 Methodology of Calculation

The procedure to assess exergy analysis has been carried out through a plaster mold factory as an example given to show the way of proceeding. However, the procedure might be achieved with resin mold factory parameters to obtain corresponding results.

Modeling of the system has been approached from a thermal point of view for every sub-process within the overall process as per Chap. 4. The efficiencies and losses are as well considered as usual in thermodynamic literature [24]. The different components of the system and considered streams have been defined as follows and shown in Figs. 10.2 and 10.3.

- Standard Power plant supplying public network (PP)-

Inlets: NG and air for combustion

Outlets: Exhausting, thermal losses and electrical power

- Boiler (BO):

Inlets: NG and air for combustion and water to be warmed up

Outlets: Exhausting, thermal losses and warm water

- Ware-dryer (WD):

Inlets: NG and air for combustion, drying air, atmospheric moisture, water from ware, sprayed water, ceramic ware, holding carts.

Outlets: Exhausting, thermal losses, water vapor, ceramic ware, holding carts

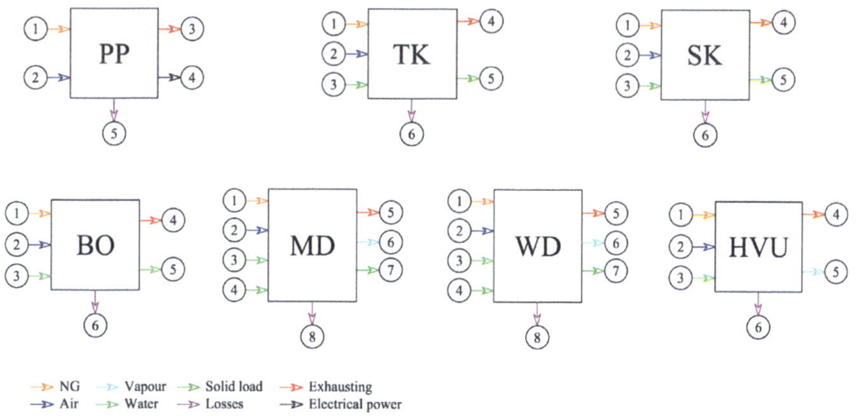

Fig. 10.2 Flowchart in a conventional factory

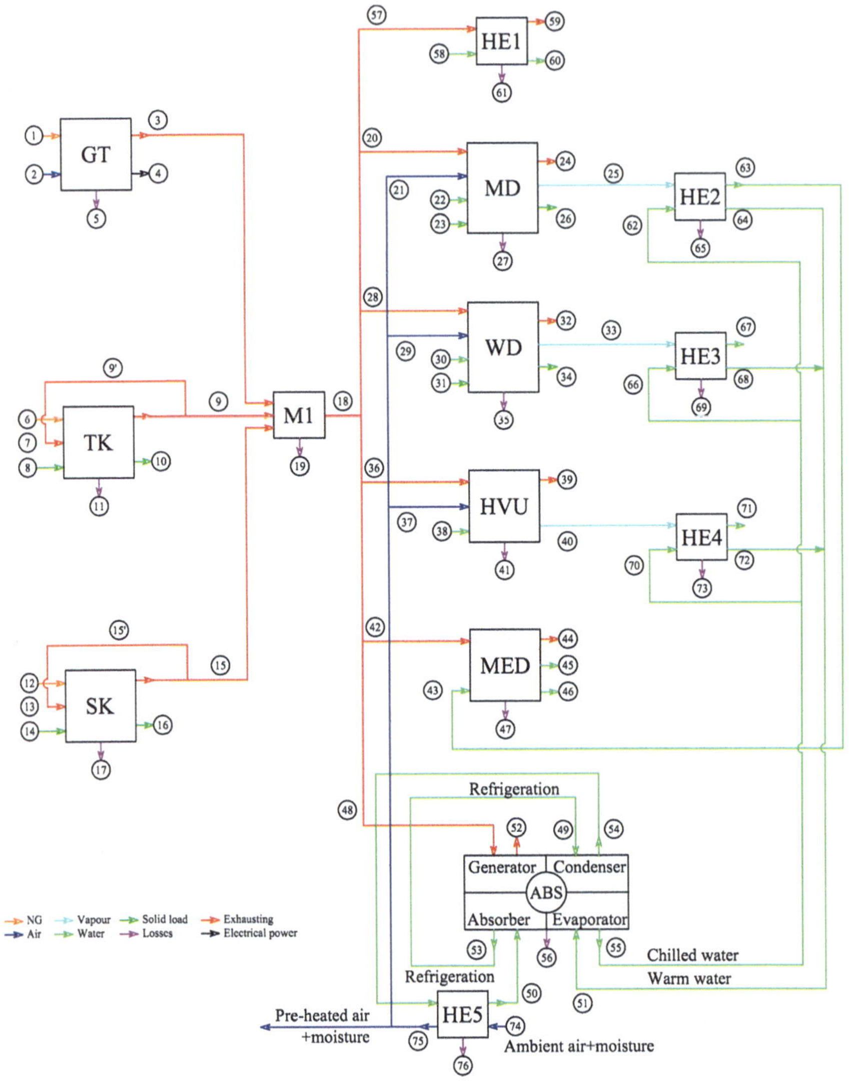

Fig. 10.3 Flow chart in an optimized factory

- Mold-dryer (WD):

Inlets: NG and air for combustion, drying air, atmospheric moisture, water from molds, plaster molds and holding supports.

Outlets: Exhausting, thermal losses, water vapor, dried plaster molds, holding supports

- Heating and ventilation unit (HVU):

Inlets: NG and air for combustion, drying air, water from ware.

Outlets: Exhausting, thermal losses, water vapor.

- Tunnel kiln. (TK):

Inlets: NG and air for combustion, diffusion, excess and cooling air, ware supports, ceramic ware, holding carts.

Outlets: Exhausting, thermal losses, ware supports, ceramic ware, holding carts.

- Shuttle kiln. (SK):

Inlets: NG and air for combustion, diffusion, injection and cooling air, ware supports, ceramic ware, holding carts

Outlets: Exhausting, thermal losses, ware supports, ceramic ware, holding carts.

- Cogeneration Gas turbine (GT):

Inlets: NG and air for combustion.

Outlets: Exhausting, thermal losses and electrical power.

- Multi-effect distiller (MED):

Inlets: recovered heat gas and water.

Outlets: Exhausting, thermal losses distillate and brine.

- Absorption machine (ABS):

Inlets: recovered heat gas and water.

Outlets: Exhausting, thermal losses, chilled water.

According to the previous models, every component of the system has been individually analyzed in order to state and calculate their energy and exergy flows by identifying every thermodynamic variable as stated in Figs. 10.2 and 10.3. These figures represent the thermal map of both configurations by means of the thermodynamic flows through the overall thermal process.

The set of mathematical formulae has been taken from [25]. This a very suitable source since it regards a single ceramic process within the tile production. This procedure is applied in the scope of this research furthering the limits of the process involving the whole factory and transferring it to the ceramic sanitary-ware industry. Corresponding equations are described as follows:

Mass balance

$$\sum_{\text{in}} \dot{m}_{\text{in}} = \sum_{\text{out}} \dot{m}_{\text{out}}. \tag{10.1}$$

Energy Balance

$$Q_{\text{in}} + \sum_{\text{in}} \dot{m}_{\text{in}} \cdot h_{\text{in}} = W_{\text{out}} + \sum_{\text{out}} \dot{m}_{\text{out}} \cdot h_{\text{out}}. \tag{10.2}$$

Exergy balance

$$Ex_f + \sum_{\text{in}} Ex_{\text{in}} = Ex_p + \sum_{\text{out}} Ex_{\text{out}} + Ex_l + Ex_d \tag{10.3}$$

$$Ex_{\text{gas}} = \dot{m}\left\{q_p\left[(T - T_0) - T_0 \ln\left(\frac{T}{T_0}\right)\right] + RT_0 \ln\left(\frac{P}{P_0}\right)\right\} \tag{10.4}$$

$$Ex_{\text{liquid,solid}} = \dot{m}\left\{q_p\left[(T - T_0) - T_0 \ln\left(\frac{T}{T_0}\right)\right] - v_m(P - P_0)\right\} \tag{10.5}$$

$$Ex_{NG}^{ch} = LHV_{NG} \cdot \left[1.033 + 0.0169\frac{n_H}{n_C} - \frac{0.0698}{n_C}\right]. \tag{10.6}$$

Efficiencies

$$\eta_{en} = \frac{\sum_{\text{useful,output}} E_i}{\sum_{\text{input}} E_i} \tag{10.7}$$

$$\eta_{en} = \frac{\sum_{\text{useful,output}} Ex_i}{\sum_{\text{input}} Ex_i}. \tag{10.8}$$

Exergoeconomic balance

$$C_{\text{in}} = C_{\text{out}} \tag{10.9}$$

$$C_{\text{in}} = \sum_{\text{in}} c_{\text{in}} \cdot Ex_{\text{in}} \tag{10.10}$$

$$C_{\text{out}} = \sum_{\text{out}} c_{\text{out}} \cdot Ex_{\text{out}} \tag{10.11}$$

$$y_d = \frac{Ex_d}{Ex_{ft}} \tag{10.12}$$

$$r = \frac{c_p - c_f}{c_f} \tag{10.13}$$

where:

$\dot{m}_{\text{in}}$	flow mass inwards (kg/s)
$\dot{m}_{\text{out}}$	flow mass outwards (kg/s)
Q_{in}	heat supplied to the system (kW)
h_{in}	specific enthalpy inwards (kJ/kg)
W_{out}	electrical power (kW)
h_{out}	specific enthalpy outwards (kJ/kg)
Ex_f	exergy of fuel (kW)
Ex_{in}	exergy inwards (kW)
Ex_p	exergy of product (kW)
Ex_{out}	exergy outwards (kW)
Ex_l	exergy of losses (kW)
Ex_d	exergy destroyed (kW)
Ex_{gas}	exergy of natural gas (kW)
q_p	specific heat capacity (kJ/kg/K)
T	temperature of the flow (K)
T_0	temperature of dead state (K)
R	ideal gas constant (J/mol/K)
P	pressure (N)
P_0	pressure of dead state (N)
$Ex_{\text{liquid,solid}}$	exergy of a liquid or solid flow (kW)
v_m	specific volume (m^3/kg)
Ex_{NG}^{ch}	specific chemical exergy of natural gas (kJ/kg)
LHV_{NG}	low heat value of natural gas (kJ/kg)
n_H	mole rate of hydrogen (mol)
n_C	mole rate of carbon (mol)
η_{en}	energy efficiency (p.u.)
η_{ex}	exergy efficiency (p.u.)
C_{in}	cost of exergy inwards (m.u.)
C_{out}	cost of exergy outwards (m.u.)
c_{in}	specific cost of exergy inwards (m.u./kW)
c_{out}	specific cost of exergy outwards (m.u./kW)
y_d	exergy destruction ratio (p.u.)
Ex_{ft}	exergy of the total fuel (kW)
r	relative cost difference of exergy (p.u.)
c_p	cost of exergy of product (m.u./kW)
c_f	specific cost of exergy of fuel (m.u./kW)
η_{el}	electrical efficiency (p.u.)
η_{th}	thermal efficiency (p.u.)
E_{NG}	energy of Natural Gas (kW)
E_f	energy of fuel (kW)
E_p	energy of product (kW)
E_{in}	energy inwards (kW)
E_{out}	energy outwards (kW)
Ex_{des}	exergy destroyed (kW)
Ex_{rec}	recoverable exergy (kW).

This set of formulae is applied to every component of both configurations, CF and OF to find out what are the balances of energy and exergy. The data and parameters of every sub-process come from commercial catalogs and technical literature regarding the machinery required for any of them. References [15, 26] are the main sources for these data. In order to produce and advanced exergetic analysis, it is required to identify the "avoidable" and "unavoidable" exergies. They allow quantifying how far a system can be optimized [5] by improving the construction design. Unavoidable exergy comes from inefficiencies due to either technical or economic limitations. Some assumptions have to be made since the total efficiency is not practically achieved. Table 10.1 shows the considerations taken into account in the scope of this research for optimal equipment versus standard or real one.

Comparing configurations, each other, they can be calculated savings when CF is transformed into OF. On the other hand, it can be seen as well, how the destroyed exergy can be decreased under best available conditions. This difference represents the "avoidable" destroyed exergy since it can be eliminated by changing technology. The residual exergy under best available conditions represents the "unavoidable" destroyed exergy, which cannot be avoided within the state of the art at current.

Table 10.1 Unavoidable conditions from efficiency differences [27]

Equipment			Real conditions	Optimal conditions
PP	Electrical efficiency	h_{el}	0.37	0.4
GT	Electrical efficiency	h_{el}	0.17	0.2
TK	Perfect combustion	Air excess	0.1	0
	Better thermal insulation	Thermal conductivity	100%	50%
	Better cooling efficiency	Air flow	100%	75%
SK	Perfect Combustion	Air excess	0.1	0
	Better thermal insulation	Thermal conductivity	100%	50%
	Better cooling efficiency	Air flow	100%	75%
BO	Thermal efficiency	h_{th}	0.9	0.95
MD	Better thermal insulation	Thermal conductivity	100%	50%
WD	Better thermal insulation	Thermal conductivity	100%	50%
HVU	Better thermal insulation	Thermal conductivity	100%	75%

10.3 Results of Exrgoeconomic Analysis

Tables 10.2 and 10.3 summarize the results of inputs–outputs for conventional and optimized factories in order to compare the increase of efficiency.

It can be seen that energy consumption and consequently CO_2 emissions get reduced by 36%. On the other hand, the water input in the conventional factory is lost by either evaporation or molds flushing. However, in the OF, most the water inlet is recovered by either distillation or condensation (except rejected brine at distiller). Moreover, there is a surplus of recovered water at the condensers coming from atmospheric and combustion moisture.

Table 10.4 shows the results for every thermodynamic variable of the flows within the system for the CF from conventional exergy analysis. It states the mass and temperature of the flow, inlet and outlet flows of energy and exergy and destroyed exergy so as the water required to be heated in the boilers to flush the molds. The inlet exergy for the CF is 8,803 kW and the outlet is 3,782 kW, hence the total destroyed exergy equals 5,021 kW. The exergy of NG is 8,480 kW.

Table 10.5 is analogous to Table 10.4 for the OF, though there are some little differences. Outlets of some sub-processes are used for inlets in other sub-processes.

Table 10.2 Summary of Inputs–Outputs for a conventional factory

Conventional factory							
Inputs	m_{IN}	E_{IN}	Ex_{IN}	Outputs	m_{OUT}	E_{OUT}	Ex_{OUT}
	kg/s	kW	kW		kg/s	kW	kW
NG	0.1748	8,482	8,741	Power		1,377	1,377
Air	22.9452	485	15	Flue gases	23.1200	5,499	1,617
Water	1.4559	378	47	Water	1.4559	1,376	222
Solids	5.4476	79	0	Solids	5.4476	263	20
				Losses	0	910	673
Total	30.0236	9,425	8,803	Total	30.0236	9,425	3,909

Table 10.3 Summary of Inputs–Outputs for an optimized factory

Optimized factory							
Inputs	m_{IN}	E_{IN}	Ex_{IN}	Outputs	m_{OUT}	E_{OUT}	Ex_{OUT}
	kg/s	kW	kW		kg/s	kW	kW
NG	0.1124	5,402	5,568	Power	0.0000	677	857
Air	110.7137	3,413	241	Flue gases	110.7669	5,655	77
Water	1.2205	129	0	Water	1.2797	257	7
Solids	5.4476	79	0	Solids	5.4476	263	36
				Losses	0.0000	2,171	943
Total	117.4942	9,022	5,809	Total	117.4942	9,022	1,920

Table 10.4 Flow rates for a CF

Equipment	Stream	i/o	Flow	m	T	E	Ex	E_in	E_out	Ex_in	Ex_out	Ex_dest	Ex_rec
				kg/s	K	kW	kW	kW	kW	kW	kW	kW	kW
PP	1	I	NG	0.0783	298	3,764	3,880	3,764		3,880			
	2	I	Air	2.3605	298	59	3	59		3			
	3	O	Exhausting	2.4389	879	2,110	812		2,110		812		812
	4	O	Power		879	1,377	1,377		1,377		1,377		
	5	O	Losses		879	335	271		335		271		
								3,823	3,823	3,883	2,460	1,422	812
TK	1	I	NG	0.0420	298	2,017	2,079	2,017		2,079			
	2	I	Air	4.8035	298	123	0	123		0			
	3	I	Ware, supports and carts	3.6666	298	50	0	50		0			
	4	O	Flue gas and air	4.8455	561	1,681	422		1,681		422		422
	5	O	Ware, supports and carts	3.6666	373	200	16		200		16		
	6	O	Losses		1,192	309	231		309		231		
								2,190	2,190	2,079	669	1,411	422
SK	1	I	NG	0.0178	298	940	969	940		969			
	2	I	Air	1.5529	298	39	0	39		0			
	3	I	Ware, supports and carts	0.7821	298	11	0	11		0			
	4	O	Flue gas and air	1.5707	755	906	324		906		324		324
	5	O	Ware, supports and carts	0.7821	350	42	4		42		4		
	6	O	Losses		755	42	25		42		25		
								990	990	969	353	616	324

(continued)

Table 10.4 (continued)

Equipment	Stream	i/o	Flow	*m*	*T*	*E*	*Ex*	*E*_in	*E*_out	*Ex*_in	*Ex*_out	*Ex*_dest	*Ex*_rec
				kg/s	K	kW	kW	kW	kW	kW	kW	kW	kW
BO	1	I	NG	0.0022	298	105	108	105		108			
	2	I	Air	0.0457	298	1	0	1		0			
	3	I	Water	0.9813	298	103	0	103		0			
	4	O	Flue gas and air	0.0479	393	19	3		19		3		
	5	O	Water	0.9813	318	185	3		185		3		
	6	O	Losses		1,799	5	4		5		4		
								209	209	108	10	98	0
MD	1	I	NG	0.0095	298	459	472	459		472			
	2	I	Air	1.3661	298	34	0	34		0			
	3	I	Water	0.1204	298	13	0	13		0			
	4	I	Molds and supports	0.1315	298	2	0	2		0			
	6	O	Flue gas and air	1.3757	333	138	13		138		13		
	7	O	Vapor from molds	0.1204	333	315	58		315		58		
	8	O	Molds and supports	0.1315	333	6	0		6		0		
	9	O	Losses		333	50	5		50		5		
								508	508	472	76	396	0
WD	1	I	NG	0.0086	298	414	427	414		427			
	2	I	Air	2.9391	298	74	0	74		0			
	3	I	Water	0.1279	298	61	13	61		13			

(continued)

Table 10.4 (continued)

Equipment	Stream	i/o	Flow	*m*	*T*	*E*	*Ex*	*E*_in	*E*_out	*Ex*_in	*Ex*_out	*Ex*_dest	*Ex*_rec
				kg/s	K	kW	kW	kW	kW	kW	kW	kW	kW
	4	I	Ware and carts	0.8674	298	15	0	15		0			
	6	O	Flue gas and air	2.9477	330	217	14		217		14		
	7	O	Vapor	0.1279	330	296	54		296		54		
	8	O	Ware and carts	0.8674	298	16	0		16		0		
	9	O	Losses		330	37			37		0		
								565	565	440	68	372	0
HVU	1	I	NG	0.0163	307	783	806	783		806			
	2	I	Air	9.8772	288	156	12	156		12			
	3	I	Water	0.2262	307	202	34	202		34			
	4	O	Flue gas and air	9.8935	307	428	30		428		30		
	5	O	Vapor	0.2262	307	580	108		580		108		
	6	O	Losses		307	132	136		132		136		
								1,140	1,140	852	274	578	0

Table 10.5 Flow rates for a OF

Equipment	Stream	i/o	Flow	*m*	*T*	*E*	*Ex*	*E*_in	*E*_out	*Ex*_in	*Ex*_out	*Ex*_dest	*Ex*_rec
				kg/s	K	kW	kW	kW	kW	kW	kW	kW	kW
GT	1	I	NG	0.0790	298	3,797	3,914	3,797		3,914			
	2	I	Air	3.3604	298	84	4	84		4			
	3	O	Exhausting	3.4395	778	2,344	829		2,344		829		829
	4	O	Power			677	857		677		857		
	5	O	Losses			860	642		860		642		
								3,881	3,881	3,918	2,327	1,590	829
TK improved	6	I	NG	0.0227	298	1,089	1,123	1,089		1,123			
	7	I	Air	4.4385	370	444	0	444		0			
	8	I	Ware, supports and carts	3.6666	298	50	0	50		0			
	9	O	Flue gas and air	4.4612	423	1,075	129		1,075		129		129
	10	O	Ware, supports and carts	3.6666	373	200	16		200		16		
	11	O	Losses		1,179	309	231		309		231		
								1,584	1,584	1,123	376	747	129
SK improved	12	I	NG	0.0107	298	515	531	515		531			
	13	I	Air	1.4189	512	352	125	352		125			
	14	I	Ware, supports and carts	0.7821	298	11	0	11		0			
	15	O	Flue gas and air	0.2698	760	795	292		795		292		292

(continued)

Table 10.5 (continued)

Equipment	Stream	i/o	Flow	m	T	E	Ex	E_in	E_out	Ex_in	Ex_out	Ex_dest	Ex_rec
				kg/s	K	kW	kW	kW	kW	kW	kW	kW	kW
	16	O	Ware, supports and carts	0.7821	350	42	4		42		4		
	17	O	Losses		760	42	25		42		25		
								878	878	656	321	335	292
M1	3	I	RH GT	3.4395	778	2,344	829	2,344		829			
	9	I	RH TK	3.7891	423	702	108	702		108			
	15	I	RH SK	0.2698	760	209	68	209		68			
	18	O	Total RH	7.4984	598	3,229	871		3,229		871		
	19	O	Losses		598	26	13		26		13		
								3,255	3,255	1,006	884	121	0
MD	20	I	RH IN	1.4206	598	539	156	539		156			
	21	I	Fresh air IN	1.0534	298	26	0	26		0			
	22	I	Water from molds	0.1204	298	13	0	13		0			
	23	I	Molds and supports	0.1315	298	2	0	2		0			
	24	O	Flue gas and air	2.4504	333	148	5		148		5		5
	25	O	Vapor	0.1441	333	376	69		376		69		
	26	O	Molds and supports	0.1315	333	6	0		6		0		
	27	O	Losses		333	50	5		50		5		
								580	580	156	79	77	5
WD	28	I	RH IN	1.2607	598	540	146	540		146			

(continued)

Table 10.5 (continued)

Equipment	Stream	i/o	Flow	*m*	*T*	*E*	*Ex*	*E*_in	*E*_out	*Ex*_in	*Ex*_out	*Ex*_dest	*Ex*_rec
				kg/s	K	kW	kW	kW	kW	kW	kW	kW	kW
	29	I	Fresh air IN	2.6335	298	106	10	106		10			
	30	I	Water from ware	0.1082	298	11	0	11		0			
	31	I	Ware and carts	0.8674	298	16	0	16		0			
	32	O	Flue gas and air	3.8360	330	218	10		218		10		10
	33	O	Vapor	0.1664	330	434	80		434		80		
	34	O	Ware and carts	0.8674	298	16	16		16		16		
	35	O	Losses		330	4	0		4		0		
								672	672	156	106	50	10
HVU	36	I	RH IN	0.1671	598	72	19	72		19			
	37	I	Fresh air IN	0.6954	288	23	3	23		3			
	38	I	Water from ware and molds	0.0105	313	2	0	2		0			
	39	O	Flue gas and air	0.8520	313	35	1		35		1		1
	40	O	Vapor	0.0211	313	54	10		54		10		
	41	O	Losses		313	8	0		8		0		
								96	96	23	12	11	1
MED	42	I	RH IN	0.9677	325	321	92	321		92			
	43	I	Water	1.0498	25	110	0	110		0			
	44	O	Flue gas and air	0.9677	44	42	1		42		1		1
	45	O	Rejected brine	0.3956	44	72	1		72		1		

(continued)

Table 10.5 (continued)

Equipment	Stream	i/o	Flow	m	T	E	Ex	E_in	E_out	Ex_in	Ex_out	Ex_dest	Ex_rec
				kg/s	K	kW	kW	kW	kW	kW	kW	kW	kW
	46	O	Distillate	0.6542	48	130	2		130		2		
	47	O	Losses		48	186	13		186		13		
								431	431	92	17	75	1
ABS	48	I	RH IN	4.9570	598	1,643	476	1,643		476			
	49	I	Water condenser IN	31.8580	306	4,446	16	4,446		16			
	50	I	Water Absorber IN	31.8580	303	3,926	5	3,926		5			
	51	I	Water evaporator IN	24.3614	285	1,249	28	1,249		28			
	52	O	Flue gas and air OUT	4.9570	380	534	53		534		53		53
	53	O	Water condenser OUT	31.8580	312	5,207	43		5,207		43		
	54	O	Water Absorber OUT	31.8580	306	4,446	16		4,446		16		
	55	O	Water evaporator OUT	24.3614	280	677	59		677		59		
	56	O	Losses		299	401	1		401		1		
								11,265	11,265	525	173	352	53
HE1	57	I	RH IN	0.3795	598	162	44	162		44			
	58	I	Cool water	0.9813	298	103	0	103		0			
	59	O	RH OUT	0.3795	328	54	7		54		7		7
	60	O	Warm water	0.9813	318	185	3		135		3		
	61	O	Losses		463	27	9		27		9		
								265	265	44	19	25	7

(continued)

Table 10.5 (continued)

Equipment	Stream	i/o	Flow	*m*	*T*	*E*	*Ex*	*E*_in	*E*_out	*Ex*_in	*Ex*_out	*Ex*_dest	*Ex*_rec
				kg/s	K	kW	kW	kW	kW	kW	kW	kW	kW
HE2	25	I	Vapor	0.0480	333	125	23	125		23			
	62	I	Cooling water	5.0241	280	140	12	140		12			
	63	O	Condensate	0.0480	333	12	0		12		0		0
	64	O	Warm water	5.0241	285	258	6		258		6		
	65	O	Losses		333	−5	0		−5		0		
								265	265	35	6	30	0
HE3	33	I	Vapor	0.1605	329	418	77	418		77			
	66	I	Cooling water	17.0561	280	474	42	474		42			
	67	O	Condensate	0.1605	329	38	1		38		1		
	68	O	Warm water	17.0561	285	875	20		875		20		
	69	O	Losses		329	−20	−2		−20		−2		
								892	892	119	19	100	0
HE4	40	I	Vapor	0.0214	327	56	10	56		10			
	70	I	Cooling water	2.2812	280	63	6	63		6			
	71	O	Condensate	0.0214	327	5	0		5		0		0
	72	O	Warm water	2.2812	285	117	3		117		3		
	73	O	Losses		327	−3	0		−3		0		
								119	119	16	3	13	0
HE5	54	I	Warm water IN	31.8580	312	5,207	43	5,207		43			

(continued)

Table 10.5 (continued)

Equipment	Stream	i/o	Flow	*m*	*T*	*E*	*Ex*	*E*_in	*E*_out	*Ex*_in	*Ex*_out	*Ex*_dest	*Ex*_rec
				kg/s	K	kW	kW	kW	kW	kW	kW	kW	kW
	74	I	Fresh air IN	101.4959	298	2,533	112	2,533		112			
	50	O	Cooled water OUT	31.8580	303	3,926	5		3,926		5		5
	75	O	Warm air OUT	101.4959	308	3,527	128		3,527		128		
	76	O	Losses		304	286	6		286		6		
								7,739	7,739	156	138	17	5

Hence, the overall factory inlets and outlets do not equal the sum of them for every sub-process. In this case, global inlet exergy results 5,809 kW and the outlet 4,584 kW. Destroyed exergy is 1,225 kW. Calculated as sum of sub-process, the sum of inlets equals 8,024 kW, and outlets equal 4,480 kW. The exergy of NG is 5,568 kW. The flushing or evaporated water appears as well in this table, but at this time, it results already heated after the distillation process or condensed after leaving dryers and HVU. The total amount of freshwater has been decreased but the losses from brine evacuation.

Table 10.6 shows the results of exergetic analysis applied to a conventional factory. First, it is considered the current state of the art standardized for the facilities. Then it is assumed some degree of improvement of efficiency of the machinery in order to find how far destroyed exergy may be cut by "avoidable" and "unavoidable" exergies. This table confronted with Table 10.4 allows determining the achievable level of optimization for CF when best available technology is applied to design it and to obtain the advanced exergetic analysis. In this case, the avoidable exergy for the CF results 959 kW.

Table 10.7 shows the exergoeconomic analysis for both standard and optimal efficiencies of equipment in an optimized factory. The complete set of exergoeconomic variables for both factories is presented in this table; energies, exergies, efficiencies and cost of exergies for every component. The cost per unit of energy flow is set to be 1 m.u./kW for the NG, so all costs within Table 10.7 are referred to the real cost of the NG.

Figure 5.8 summarizes the thermal and water consumption of both, CF and OF. The main differences lay in the heat recovery from kilns exhausting and the water to flush molds, which is not recovered in a CF. The water saving proposed in this research lays on re-use it through a distillation process. There are similar approaches within industrial field. In [28], the way to reduce the water consumption is based on re-cycling the wasted brine of the R.O. (reverse osmosis) process. Gas consumption is raised up due to the GT. On the other hand, it is cut down due to heat recovery from kilns. Balance is positive from the reduction point of view. Besides, the electrical power is not supplied at public price, which derives an important economic saving, so as the water reduction that involves an environmental improvement as well.

Advance exergy analysis has been used to optimize the CF finding the flows of exergy through the different elements in order to re-conduct them taking advantage of residual exergy at the kilns exhausting. From Table 10.4, it can be concluded that 422 kW of exergy from exhausting of the kilns and 834 from power plant could be profited to supply boilers, dryers and HVU. Half of the exergy results destroyed through the global process. From Tables 10.4 and 10.6, it is deduced that 959 kW of exergy are avoidable in a CF by difference between the destroyed exergies in a real or unavoidable condition. Regarding efficiency, it can be seen how equipment behavior is improved when fed from recovered heat, so as the global process efficiency gets higher. The analysis is repeated for the OF. As a consequence of this fact, the rate of destroyed exergy and the exergetic cost of product are lowered.

Compared overall factory results for both configurations are shown in Table 10.8.

Table 10.6 Assessment of avoidable and unavoidable destroyed exergy in a conventional factory

Real Conditions

Equipment	E_f	Energy			Exergy								Exergoeconomy				
		E_p	E_{eff}	Ex_{in}	Ex_{out}	Ex_f	Ex_p	Ex_d	Ex_{eff}	y_d	Ex_{d}_av	C_f	C_p	c_f	c_p	r	E_f
		kW	kW	p.u.	kW	kW	kW	kW	kW	p.u.	p.u.	kW	kW	kW	m.u./kW	m.u./kW	p.u.
PP	3,764	3,764	1,377	0.3659	3,883	2,460	3,880	1,377	1,422	0.3550	0.1677	331	3,880	3,880	1.00	2.82	1.8169
TK	2,017	2,017	529	0.2621	2,079	669	2,079	338	1,411	0.1624	0.1664	414	2,079	2,079	1.00	6.16	5.1582
SK	940	940	154	0.1642	969	353	875	99	616	0.1133	0.0726	128	875	875	1.00	8.82	7.8240
BO	105	105	82	0.7834	108	10	108	3	98	0.0248	0.0116	6	108	108	1.00	40.33	39.3270
MD	458	458	302	0.6592	472	76	472	58	396	0.1223	0.0467	26	472	472	1.00	8.18	7.1793
WD	414	414	230	0.5553	440	73	427	44	367	0.1038	0.0433	19	427	427	1.00	9.63	8.6299
HVU	782	782	376	0.4812	852	142	806	74	711	0.0920	0.0838	36	806	806	1.00	10.87	9.8682
Overall	8,480	8,480	3,051	0.3597	8,803	3,782	8,647	1,993	5,021	0.2305	0.5921	959	8,647	8,647	1.00	4.34	3.3385

Unavoidable Conditions = Optimal

Equipment		Energy			Exergy								Exergoeconomy				
		E_f	E_p	E_{eff}	Ex_{in}	Ex_{out}	Ex_f	Ex_p	Ex_d	Ex_{eff}	y_d	Ex_{d}_av	C_f	C_p	c_f	c_p	r
		kW	kW	p.u.	kW	kW	kW	kW	kW	p.u.	p.u.	kW	kW	kW	m.u./kW	m.u./kW	p.u.
PP	3,443	3,443	1,377	0.4000	3,552	2,460	3,550	1,377	1,092	0.3880	0.1287	–	3,550	3,550	1.00	2.58	1.5770
TK	1,616	1,616	529	0.3271	1,666	669	1,666	338	997	0.2027	0.1176	–	1,666	1,666	1.00	4.93	3.9336
SK	815	815	154	0.1893	841	353	759	99	488	0.1306	0.0575	–	759	759	1.00	7.66	6.6563
BO	99	99	82	0.8269	102	10	102	3	92	0.0262	0.0109	–	102	102	1.00	38.20	37.2045
MD	433	433	302	0.6972	447	76	447	58	371	0.1293	0.0437	–	447	447	1.00	7.73	6.7338
WD	396	396	230	0.5809	420	73	408	44	347	0.1086	0.0410	–	408	408	1.00	9.20	8.2045
HVU	749	749	376	0.5024	816	142	772	74	675	0.0961	0.0795	–	772	772	1.00	10.41	9.4090

(continued)

Table 10.6 (continued)

Unavoidable Conditions = Optimal																	
Equipment		Energy			Exergy								Exergoeconomy				
		E_f	E_p	E_{eff}	Ex_{in}	Ex_{out}	Ex_f	Ex_p	Ex_d	Ex_{eff}	y_d	$Ex_{d\_}av$	C_f	C_p	c_f	c_p	r
		kW	kW	p.u.	kW	kW	kW	kW	kW	p.u.	p.u.	kW	kW	kW	m.u./kW	m.u./kW	p.u.
Overall	7,552	7,552	3,051	0.4039	7,844	3,782	7,703	1,993	4,061	0.2587	0.4790	–	7,703	7,703	1.00	3.86	2.8649

Table 10.7 Assessment of avoidable and unavoidable destroyed exergy in an optimized factory

Real Conditions																	
Equipment	Energy				Exergy								Exergoeconomy				
	E_{NG}	E_f	E_p	E_{eff}	Ex_{in}	Ex_{out}	Ex_f	Ex_p	Ex_d	Ex_{eff}	y_d	Ex_{d_av}	C_f	C_p	c_f	c_p	r
	kW	kW	kW	p.u.	kW	kW	kW	kW	kW	p.u.	p.u.	kW	kW	kW	m.u./kW	m.u./kW	p.u.
GT	3,797	3,797	677	0.1783	3,918	2,148	3,914	677	1,770	0.1730	0.4731	425	3,914	3,914	1.00	5.78	4.7814
TK	1,089	1,089	529	0.4853	1,251	525	1,123	338	726	0.3007	0.1939	460	1,123	1,123	1.00	3.33	2.3257
SK	515	515	154	0.2995	656	321	531	99	335	0.1866	0.0896	158	531	531	1.00	5.36	4.3589
M1	0	6,012	5,957	0.9908	1,993	1,848	1,993	1,818	145	0.9120	0.0388	0	5,568	5,568	3.06	3.36	0.0965
MD	0	392	302	0.7700	156	79	142	58	77	0.4062	0.0206	10	478	478	3.36	8.27	1.4621
WD	0	359	271	0.7539	156	91	122	52	66	0.4258	0.0175	1	410	410	3.36	7.89	1.3483
HVU	0	50	25	0.5009	23	12	16	5	11	0.3051	0.0030	1	55	55	3.36	11.01	2.2776
MED	0	278	244	0.8777	92	17	91	32	75	0.3495	0.0200	9	307	307	3.36	9.61	1.8614
ABS	0	1,109	572	0.5158	525	173	423	31	352	0.0739	0.0940	52	1,420	1,420	3.36	45.42	12.5233
HE1	0	162	103	0.6334	44	19	37	3	25	0.0722	0.0067	2	125	125	3.36	46.53	12.8532
HE2	0	118	113	0.9611	35	6	23	23	30	0.9792	0.0079	0	192	192	8.27	8.45	0.0213
HE3	0	401	380	0.9500	119	19	78	76	100	0.9758	0.0267	0	615	615	7.89	8.08	0.0248
HE4	0	54	51	0.9500	16	3	10	10	13	0.9774	0.0036	0	115	115	11.01	11.26	0.0232
HE5	0	1,281	994	0.7766	156	138	39	16	17	0.4115	0.0046	8	1,764	1,764	45.42	110.38	1.4301
Overall	5,402	5,402	2,850	0.5276	5,809	4,584	5,402	2,305	1,225	0.4268	0.3275	1,127	5,402	5,402	1.00	2.34	1.3433

(continued)

Table 10.7 (continued)

Unavoidable Conditions = Optimal																	
Equipment	Energy				Exergy								Exergoeconomy				
	E_{NG}	E_f	E_p	E_{eff}	Ex_{in}	Ex_{out}	Ex_f	Ex_p	Ex_d	Ex_{eff}	y_d	Ex_{d_av}	C_f	C_p	c_f	c_p	r
	kW	kW	kW	p.u.	kW	kW	kW	kW	kW	p.u.	p.u.	kW	kW	kW	m.u./kW	m.u./kW	p.u.
GT	3,385	3,385	677	0.2000	3,493	2,148	3,489	677	1,345	0.1940	0.0861	–	3,489	3,489	1.00	5.15	4.1540
TK	688	688	529	0.7681	790	525	710	338	265	0.4759	0.0170	–	710	710	1.00	2.10	1.1012
SK	391	391	154	0.3948	498	321	403	99	177	0.2460	0.0113	–	403	403	1.00	4.07	3.0655
M1	0	6,012	5,957	0.9908	1,993	1,848	1,993	1,818	145	0.9120	0.0093	–	4,602	4,602	2.53	2.78	0.0965
MD	0	367	302	0.8224	146	79	133	58	67	0.4338	0.0043	–	370	370	2.78	6.40	1.3054
WD	0	357	271	0.7585	155	91	121	52	65	0.4285	0.0041	–	337	337	2.78	6.48	1.3339
HVU	0	49	25	0.5202	22	12	16	5	10	0.3169	0.0007	–	44	44	2.78	8.76	2.1558
MED	0	250	244	0.9752	83	17	82	32	66	0.3883	0.0042	–	228	228	2.78	7.15	1.5752
ABS	0	998	572	0.5731	472	173	381	31	299	0.0822	0.0192	–	1,056	1,056	2.78	33.78	11.1709
HE1	0	154	103	0.6686	42	19	35	3	23	0.0762	0.0015	–	98	98	2.78	36.43	12.1241
HE2	0	118	113	0.9611	35	6	23	23	30	0.9792	0.0019	–	149	149	6.40	6.54	0.0213
HE3	0	401	380	0.9500	119	19	78	76	100	0.9758	0.0064	–	505	505	6.48	6.64	0.0248
HE4	0	54	51	0.9500	16	3	10	10	13	0.9774	0.0009	–	91	91	8.76	8.96	0.0232
HE5	0	1,217	994	0.8175	148	138	37	16	9	0.4332	0.0006	–	1,247	1,247	33.78	78.00	1.3086
Overall	4,464	4,464	2,321	0.5200	4,801	3,733	4,464	1,878	1,067	0.4206	0.2853	–	4,464	4,464	1.00	2.38	1.3776

Table 10.8 Exergoeconomic comparison conventional versus optimized factory

Exergoeconomy summary					
Overall factory		*CF*	$CF_{UN\ \mathrm{conditions}}$	*OF*	$OF_{UN\ \mathrm{conditions}}$
E_f	kW	8,480	7,552	5,402	4,464
E_p	kW	3,051	3,051	2,850	2,321
E_{eff}	p.u.	0.3597	0.4039	0.5276	0.5200
Ex_{in}	kW	8,803	7,844	5,809	4,801
Ex_{out}	kW	3,782	3,782	4,584	3,733
Ex_f	kW	8,647	7,703	5,402	4,464
Ex_p	kW	1,993	1,993	2,305	1,878
Ex_d	kW	5,021	4,061	1,225	1,067
Ex_{eff}	p.u.	0.2305	0.2587	0.4268	0.4206
y_d	p.u.	0.5806	0.5273	0.2269	0.2391
Ex_{d_av}	kW	959	–	1,127	–
C_f	kW	8,647	15,406	5,402	8,928
C_p	kW	8,647	15,406	5,402	8,928
c_f	m.u./kW	1.0000	2.0000	1.0000	2.0000
c_p	m.u./kW	4.3385	7.7297	2.3433	4.7552
r	p.u.	3.3385	2.8649	1.3433	1.3776

From Table 10.8, it can be deduced that NG consumption has been decreased over 36%. The exergetic efficiency rises from 0.36 up to 0.52. The destroyed exergy decreases 75% for standard conditions between CF and OP. The cost per unit of exergy flow decreases 40%. On the other hand, 74% of inlet water is reused as seen in Table 10.9.

As stated in the introduction section, the study has been based on the factory concept from [21] where 5 • 10^5 pieces are produced per year on a regular basis of labor days. Also, in the introduction, they have been mentioned some figures of

Table 10.9 Water balance in an OF

Water balance					
Inlet	kg/s	Outlet	kg/s	Recovered	kg/s
Molds MD	0.1204	Rejected brine distiller	0.3956		
Ware WD	0.1082	Distillate	0.6542	Distillate	0.6542
Molds and Ware HVU	0.0105	Condensate HE2	0.0480	Condensate HE2	0.0480
HE1 for flushing molds	0.9813	Condensate HE3	0.1605	Condensate HE3	0.1605
		Condensate HE4	0.0214	Condensate HE4	0.0214
Total	1.2205	Total	1.2797	Total	0.8841

standard consumptions within the sanitary-ware industry. Calculations have proved these surveyed data via analysis.

From Table 10.8, the theoretical thermal power required to produce ware is 8,480 kW. This means

$$\frac{8,480\ \text{kW}\cdot 10^{-3}\frac{\text{MW}}{\text{kW}}\cdot 232\frac{\text{days}}{\text{year}}\cdot 24\frac{\text{hours}}{\text{day}}\cdot 3,600\frac{\text{s}}{\text{hour}}}{5\cdot 10^{5}\frac{\text{pieces}}{\text{year}}\cdot 20\frac{\text{kg}}{\text{piece}}} = 17\frac{\text{MW}}{\text{kg}}. \tag{10.14}$$

The current standard is about 21.87 MW/kg (Table 1.1).

For a standard tunnel kiln, it was stated in Sect. 10.1, a consumption of 4.2–6.7 MJ/kg. From Table 10.4, NG power for tunnel kiln results 2,020 kW and the specific consumption results:

$$\frac{2,017\ \text{kW}\cdot 10^{-3}\frac{\text{MW}}{\text{kW}}\cdot 232\frac{\text{days}}{\text{year}}\cdot 24\frac{\text{hours}}{\text{day}}\cdot 3,600\frac{\text{s}}{\text{hour}}}{5\cdot 10^{5}\frac{\text{pieces}}{\text{year}}\cdot 20\frac{\text{kg}}{\text{piece}}} = 4.04\frac{\text{MW}}{\text{kg}}. \tag{10.15}$$

For a standard shuttle kiln, it was stated, as well in Sect. 10.1, a consumption of 8.3–10.4 MJ/kg. In this regard, it has to be remarked that only defective parts are to be fired. It has to be assumed a percentage of ware to be repaired. Let us assume a 15% and a duty cycle of 15 h/day. From Table 10.4, NG power for shuttle kiln results in 940 kW and the specific consumption results:

$$\frac{940\ \text{kW}\cdot 10^{-3}\frac{\text{MW}}{\text{kW}}\cdot 232\frac{\text{days}}{\text{year}}\cdot 15\frac{\text{hours}}{\text{day}}\cdot 3,600\frac{\text{s}}{\text{hour}}}{0.15\cdot 5\cdot 10^{5}\frac{\text{pieces}}{\text{year}}\cdot 20\frac{\text{kg}}{\text{piece}}} = 7.85\frac{\text{MW}}{\text{kg}}. \tag{10.16}$$

For a standard dryer, Sect. 10.1 stated a consumption of 0.3–1.4 MJ/kg. From Table 10.4, NG power for dryer results 414 kW and the specific consumption results:

$$\frac{414\ \text{kW}\cdot 10^{-3}\frac{\text{MW}}{\text{kW}}\cdot 232\frac{\text{days}}{\text{year}}\cdot 24\frac{\text{hours}}{\text{day}}\cdot 3,600\frac{\text{s}}{\text{hour}}}{5\cdot 10^{5}\frac{\text{pieces}}{\text{year}}\cdot 20\frac{\text{kg}}{\text{piece}}} = 0.83\frac{\text{MW}}{\text{kg}}. \tag{10.17}$$

The research has found that 2,017 kW of natural gas power is required to heat the ware up to 1,250 °C in order to allow the sintering and chemical bonds. Only 529 kW are required to heat the ware up. Hence, the efficiency of the kiln results 26% when the ware is heated. This value gets reduced when it is cooled down. It is out of the scope of this study to calculate the chemical exergetic balance within the ceramic material to find out the real final efficiency of the kiln regarding the final remaining exergy inside the ceramic ware as stated in [121] by 15%. On the other hand, the stacks of the tunnel kiln provide the following thermal power as per Table 10.4:

$$\frac{1,681\text{kW}(at610\text{K})}{2,017\text{kW}} = 83\%. \tag{10.18}$$

The recoverable fraction depends on the thermal level of the heat exchange or heat usage of it. Assuming a thermal level of 373 K, the recoverable thermal power would be approximated by:

$$\frac{1,681\text{kW}\frac{(610-373)\text{K}}{(610-273)\text{K}}}{2,017\text{kW}} = 58.6\%. \tag{10.19}$$

And destroyed exergy has resulted from Table 10.6

$$\frac{2,079\text{ kW} - 669\text{ kW}}{2,079\text{ kW}} = 68\%. \tag{10.20}$$

These values are coherent with 61% and 83% sated in [121].

On the other hand, preheating of inlet airs for the tunnel kiln reduces the NG power from 2,017 to 1,089 kW. This means a 46% of gas saving. Caglayan stated in [122] reductions values of 4.7% (at 421 K) and 8–10% (at 473–523 K). From Figs. 6.8 and 6.9, it can be deduced that thermal level of preheating in this case study is about 900 K (based on Fig. 4.21). If lineal trend is assumed, the savings of gas for this temperature would be:

$$4.7\% + \frac{(10-4.7)\%}{(523-421)\text{K}} \cdot (900-421)\text{K} = 29.59\%. \tag{10.21}$$

From these calculations, it seems that correlation may be polynomial or exponential rather than lineal. It has to be said that this great reduction of consumption is not only produced by pre-heating but also, because of induced draught.

Regarding CO_2 emissions, Table 10.10, provides the difference between both configurations, stating that one third of the released CO_2 may be avoided.

Table 10.10 CO_2 emissions balance

CO_2 Balance					
CF	kg/s	t/year	OF	kg/s	t/year
PP	0.2154	4,318	GT	0.2173	4,356
TK	0.1156	2,317	TK	0.0624	1,251
SK	0.0490	983	SK	0.0295	592
BO	0.0060	120	M1		
MD	0.0088	175	MD		
WD	0.0237	475	WD		
HVU	0.0448	898	HVU		
Total	0.4633	9,286	TOTAL	0.3092	6,198

10.4 Major Conclusions of Exergoeconomic Analysis

This case study has demonstrated how a CF for ceramic sanitary-ware can become more efficient by applying the principles of thermal engineering from a thermodynamic perspective analyzing heat transfers and the quality of the thermal energy as well as water.

Primary energy gets reduced 36% so as the CO_2 emissions and 18,062 m^3 of water per year are saved. This reduction implies almost 3,088 t of emitted CO_2 less per year.

On the other hand, the existing data regarding the subject have been proved to fit the theoretical calculation. That proves the model is appropriated to the targeted analysis of the research.

Regarding water saving, this research has found and quantified the surplus of water that can be obtained from air moisture and combustion products after condensation.

The particular conditions of resource intensity in the sanitary-ware industry provide an optimal condition to approach the water-energy nexus where in theory, the water could be reused in a rate of 74% in an industrial field, which is currently a great consumer of both, energy and water.

Exergetic analysis has been carried out to better understand the potential of saving and to identify and quantify the equipment more likely to reduce energy. In this regard, this study has contributed to extend the work done in [64] and [18, 21] by applying the advanced exergetic analysis to the modeled factories and quantifying their savings of primary energy and water.

References

1. Cuviella- Suárez C, Peiró JJB, Borge-Diez D, Colmenar-Santos A (2019) Exergoeconomics in the sanitary-ware industry to reduce energy and water consumptions. AIP Conf Proc 2190(1):020061. https://doi.org/10.1063/1.5138547
2. Ciacco EFS, Rocha JR, Coutinho AR (2017) The energy consumption in the ceramic tile industry in Brazil. Appl Therm Eng 113:1283–1289. https://doi.org/10.1016/j.applthermaleng.2016.11.068
3. Scott CA, Sugg ZP (2015) Global energy development and climate-induced water scarcity—physical limits, sectoral constraints, and policy imperatives. Energies 8(8):8211–8225. https://doi.org/10.3390/en8088211
4. Monfort E, Mezquita A, Vaquer E, Celades I, Sanfelix V, Escrig A (2014) 8.05 - ceramic manufacturing processes: energy, environmental, and occupational health issues. In: Hashmi S, Batalha GF, Van Tyne CJ, Yilbas B (eds) Comprehensive materials processing. Elsevier, Oxford, pp 71–102
5. Vuckovic G, Vukic M, Stojilkovic M (2012) Avoidable and unavoidable exergy destruction and exergoeconomic evaluation of the thermal processes in a real industrial plant. Therm Sci 16(Suppl. 2):S433–S446. https://doi.org/10.2298/TSCI120503181V
6. Dincer I, Rosen MA, Al-Zareer M (2018) 5.5 exergy management. In: Dincer I (ed) Comprehensive energy systems. Elsevier, Oxford, pp 166–201

7. Kita H, Himoto I, Yamashita S (2016) Exergetic aspects of green ceramic processing. In: Singh M, Ohji T, Asthana R (eds) Green and sustainable manufacturing of advanced material. Elsevier, Oxford, pp 77–98
8. Ferrer S, Mezquita A, Aguilella VM, Monfort E (2019) Beyond the energy balance: exergy analysis of an industrial roller kiln firing porcelain tiles. Appl Therm Eng 150:1002–1015. https://doi.org/10.1016/j.applthermaleng.2019.01.052
9. Caglayan H, Caliskan H (2018) Investigation of the energy recovery in the burners of the ceramic factory kiln. Energy Proc 144:118–124. https://doi.org/10.1016/j.egypro.2018.06.016
10. Gomez RS et al (2019) Natural gas intermittent kiln for the ceramic industry: a transient thermal analysis. Energies 12(8):1568. https://doi.org/10.3390/en12081568
11. Ros-Dosdá T, Fullana-i-Palmer P, Mezquita A, Masoni P, Monfort E (2018) How can the European ceramic tile industry meet the EU's low-carbon targets? A life cycle perspective. J Clean Prod 199:554–564. https://doi.org/10.1016/j.jclepro.2018.07.176
12. Mezquita A, Monfort E, Ferrer S, Gabaldón-Estevan D (2017) How to reduce energy and water consumption in the preparation of raw materials for ceramic tile manufacturing: dry versus wet route. J Clean Prod. https://doi.org/10.1016/j.jclepro.2017.04.082
13. LIFE_ENVIP.pdf (2012). http://ec.europa.eu/environment/life/project/Projects/index.cfm?fuseaction=search.dspPage&n_proj_id=4630&docType=pdf. Accessed 02 Dec 2016
14. Caglayan H, Şöhret Y, Caliskan H (2018) Thermo-ecologic evaluation of a spray dryer for ceramic industry. Energy Proc 144:164–169. https://doi.org/10.1016/j.egypro.2018.06.022
15. Reference Document on Best Available Techniques in the Ceramic Manufacturing Industry. European Commission, August 2007 (2007). http://eippcb.jrc.ec.europa.eu/reference/BREF/cer_bref_0807.pdf. Accessed 02 Dec 2016
16. Cuviella-Suárez C, Colmenar-Santos A, Borge-Diez D, López-Rey Á (2018) Management tool to optimize energy and water consumption in the sanitary-ware industry. J Clean Prod 197, Part 1:280–296. https://doi.org/10.1016/j.jclepro.2018.06.195
17. Bi R, Chen C, Tang J, Jia X, Xiang S (2019) Two-level optimization model for water consumption based on water prices in eco-industrial parks. Resour Conserv Recycl 146:308–315. https://doi.org/10.1016/j.resconrec.2019.04.004
18. Cuviella-Suárez C, Colmenar-Santos A, Borge-Diez D, López-Rey Á (2019) Heat recovery in sanitary-ware industry applied to water and energy saving by multi-effect distillation. J Clean Prod 213:1322–1336. https://doi.org/10.1016/j.jclepro.2018.12.269
19. Tonn B, Frymier PD, Stiefel D, Skinner LS, Suraweera N, Tuck R (2014) Toward an infinitely reusable, recyclable, and renewable industrial ecosystem. J Clean Prod 66:392–406. https://doi.org/10.1016/j.jclepro.2013.11.008
20. da Silva AC, Méxas MP, Quelhas OLG (2017) Restrictive factors in implementation of clean technologies in red ceramic industries. J Clean Prod 168:441–451. https://doi.org/10.1016/j.jclepro.2017.09.086
21. van Vugt M, Moye A, Sivakumar S (2019) Computational modelling approaches to meditation research: why should we care? Curr Opin Psychol 28:49–53. https://doi.org/10.1016/j.copsyc.2018.10.011
22. Brailsford SC, Eldabi T, Kunc M, Mustafee N, Osorio AF (2019) Hybrid simulation modelling in operational research: a state-of-the-art review. Eur J Oper Res 278(3):721–737. https://doi.org/10.1016/j.ejor.2018.10.025
23. Amini SH, Remmerswaal JAM, Castro MB, Reuter MA (2007) Quantifying the quality loss and resource efficiency of recycling by means of exergy analysis. J Clean Prod 15(10):907–913. https://doi.org/10.1016/j.jclepro.2006.01.010
24. Rosen MA (2002) Clarifying thermodynamic efficiencies and losses via exergy. Exergy Int J 2(1):3–5. https://doi.org/10.1016/S1164-0235(01)00054-1
25. Caglayan H, Caliskan H (2018) Energy, exergy and sustainability assessments of a cogeneration system for ceramic industry. Appl Therm Eng 136:504–515. https://doi.org/10.1016/j.applthermaleng.2018.02.064
26. Ceramic Technology Sanitary-ware, 2010th edn. SACMI IMOLA S.C. (2010)

27. Tsatsaronis G, Cziesla F, Gao Z (2003) Avoidable thermodynamic inefficiencies and costs in energy conversion systems. Part 1: methodology. In: Proceedings of ECOS 2003, January 2003, vol 2, pp 809–814
28. Ahmadi M, Baniasadi E, Ahmadikia H (2017) Process modeling and performance analysis of a productive water recovery system. Appl Therm Eng 112:100–110. https://doi.org/10.1016/j.applthermaleng.2016.10.067

Chapter 11
Practical Implementation of Selected Configurations

As a summary of the optimization measures to improve the performance of the factory, it should be analyzed the economic impact with an assessment of the required investments. In order to evaluate the impact of the investment, they will be compared the cost of the investments and the achieved savings [1]. Savings of resources calculated in Chap. 6 have to be translated into currency by means of the corresponding unitary price. Price of electricity is assumed to be 100 €/MWh [2], gas price 25 €/MWh [3] and water price, 1.5 €/m^3 [4] (Figs. 11.1 and 11.2).

A prior elemental analysis is made by means of a simple pay-back quantification

$$r = \frac{I}{S} \tag{11.1}$$

where:

r: simple pay-back (years)

I: investment (€)

S: economic savings (€).

$$S = RS \cdot c_e \tag{11.2}$$

where:

RS: resource savings (kWh)

c_e: resource price (€/kWh) or (€/m^3).

C. Cuviella-Suárez et al., *Water and Energy Use in Sanitary-ware Manufacturing*,
Green Energy and Technology, https://doi.org/10.1007/978-3-030-72491-7_11

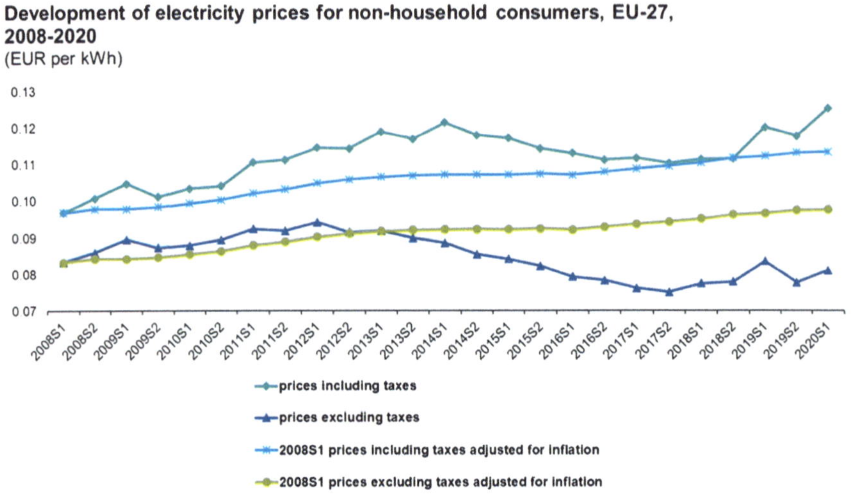

Fig. 11.1 Electricity price evolution in EU (*Source* EUROSTAT)

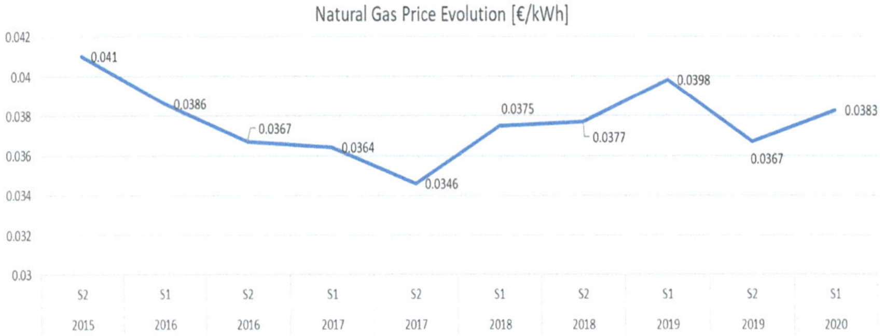

Fig. 11.2 Natural gas price evolution in EU (*Source* EUROSTAT)

11.1 Electrical General Measures

11.1.1 Efficiency of the Electrical Motors

The price of the motors is simulated with a polynomial correlation from a price list as follows:

$$P_{\text{motor}} = a_0 + a_1 \cdot N + a_2 \cdot N^2 + a_3 \cdot N^3 + a_4 \cdot N^4 + a_5 \cdot N^5 \quad (11.3)$$

where:

p_{motor}: price of the motor (€)

N: electrical power of the motor (kW)

a_i: polynomial coefficients for either IE2 or IE3 class.

Price motor polynomial coefficients			
	IE2		IE3
a0	585.262		542.7072
a1	141.6953		233.4881
a2	0.94116		−0.61477
a3	−0.00763		0.001644
a4	2.19E-05		−1.85E-07
a5	−2.02E-08		−2.20E-09

Investment approach may be considered for a new facility, so the cost of the measure would be just the difference of prices between motors. In the event that facility is already operating the investment should be applied to the most efficient motor since it has to be actually substituted.

11.1.2 Frequency Variation of the Electrical Motors

In the same line that the previous section, it is compared the price of frequency variation implementation to each of the motors that may be regulated.

11.1.3 Substitution of Conventional Lighting by LED

Assumed cost for conventional lighting is:

Cost of LED substitution	€/W	
	Fluorescence	Discharge
Conventional	0.0612	0.1240
LED	0.3750	0.4333

11.1.4 Lighting Regulation

This action is a very variable investment depending on the degree of automation and segmentation of the areas to be individually regulated. In this case study, it has be roughly approximated by 0.25% of the regulated power.

11.2 General Thermal Measures

11.2.1 Increase of Thermal Insulation

The investment of this action is based on the reinforcement of insulation panel around the chamber to be isolated; dryers or casting hall.

$$I = S \cdot C_{\text{panel}} \tag{11.4}$$

where:

S: surface to be covered by panel (m^2)

C_{panel}: specific cost of the panel implementation (€/m^2). This value has been considered 40 €/m^2.

11.2.2 Thermal and Humidity Automatic Control

It is based on a frequency variation to adjust the speed of the fans. Thus, the investment is the sum of the cost of converters, wiring electrical boards, sensors and actuators, etc.

11.2.3 Load Lightening by SiC

Comparing the difference on load weighs between cordierite and SiC calculated in Sect. 6.9.5 and the prices of the materials, it can be computed the investment value. Cordierite cost is considered 5€/kg and Sic has been set as 15 €/kg.

11.2.4 Heating Recovery from Compressor Refrigeration

The heat of the compressor chiller has to be transferred to a distribution circuit throughout a heat exchanger. This configuration may vary much depending on the system, equipment, distances, etc. In this case study, the specific cost of implementation has been roughly approximated by 150 €/kW.

11.3 Exclusive Measures

11.3.1 Heating Recovery

Recovered heat from kilns has to be blown into metal ducts to be distributed through the different consumers. These ducts must be properly insulated in order to minimize heat losses. On the other hand, miscellaneous equipment has to be considered; filters, heat exchangers, fans for high temperature, supporting elements, etc. A good approximation of this investment may be 750 €/kW.

11.3.2 Induced Draught in Tunnel Kiln

As explained in Sect. 6.9.1, it is based on a proper control of the exhausting fan of the tunnel kiln. It is achieved by a frequency converter with all related sensor/actuator equipment so as the programmed controller to meet the thermal curve inside the kiln with the minimal amount of fresh air inwards. The cost may vary depending on the degree of automation. For the purpose of this case study, it has been considered a cost of 150,000 €.

11.3.3 Pre-heating of Kiln Airs

As explained in Sects. 7.3.3 and 7.3.4, the heat must be transferred from cooling section towards firing zone. Thus, a network of pipes has to be set so as to propelling fans. Because of the high temperature of some sections, ceramic or refractory elements must be considered. The cost for the purpose of this book has been set as 200,000 € for the tunnel kiln and 150,000 for the shuttle kiln.

11.3.4 Absorption Machine

Specific cost for these types of systems can be assumed to be 575 €/kW

11.3.5 MED Distiller

Specific cost for these type of systems can be assumed to be 2,500 €/m^3·day.

11.3.6 Photovoltaic Plant

Specific cost for these type of systems can be assumed to be 600 €/kW.

11.3.7 Cogeneration Plant

Specific cost for these type of systems can be assumed to be 2,500 €/kW

11.4 Summary of Investment

After analyzing the required investment to carry out the saving measures and the economic impact, it can be built Table 11.1, where the savings and investments are shown, so as the corresponding pay-back (Table 11.2).

In this case, the payback periods are very long because of the relative low price of water. These periods may decrease drastically as the price rises up induced by regulations, lack of freshwater, etc.

As a prior way to have an idea on the viability of investments, this pay-back analysis is enough, but in order to go deeper into analysis, a present value of the investments must be carried out for a more accurate calculation.

Financial parameters for this analysis are shown (Table 11.3).

It can be defined the levelized cost of energy as:

$$LCE = \frac{LC_{\text{Inv}} + LC_{O\&M} + LC_{\text{fuel}}}{E_{\text{saved}}} \tag{11.5}$$

where:

LCE: levelized cost of energy (€/kWh)

LC_{Inv}: levelized cost of investment (€/kWh)

$LC_{O\&M}$: levelized cost of operation and maintenance (€/kWh)

LC_{fuel}: levelized cost of fuel (€/kWh).

It represents the specific cost of the energy saving by each investment. The less this value, the most convenience of this investment.

The levelized cost of the investment can be calculated by:

$$LC_{\text{Inv}} = CRF \cdot \text{Inv} \tag{11.6}$$

$$CRF = \frac{i_{\text{eff}} \cdot (1 + i_{\text{eff}})^n}{(1 + i_{\text{eff}})^n - 1} \tag{11.7}$$

Table 11.1 Summary of investments to optimize a factory from a pay-back perspective

		PMF			RMF		
		Savings	Investment	Pay Back	Savings	Investment	Pay Back
		kWh/year	€	years	kWh/year	€	Years
General electrical measures	Savings by motor efficiency	201,922	1,417,710	70.21	141,303	992,096	70.21
	Savings by frequency converters	1,609,294	1,035,251	6.43	1,273,461	819,211	6.43
	Savings by LED lighting	79,680	2,622	0.33	63,410	2,087	0.33
	Savings by lighting regulation	30,886	21,896	7.09	24,639	17,468	7.09
General thermal measures	Increase of thermal insulation of the mold dryer	138,981	15,840	4.56	–	–	–
	Increase of thermal insulation of the casting hall	255,100	308,040	48.30	87,786	184,824	84.22
	Thermal and humidity control of the HVU's	343,221	35,000	4.08	–	–	–
	Increase of thermal insulation of the ware dryer	101,809	13,300	5.23	101,809	13,300	5.23
	Thermal and humidity control of the ware dryer	144,142	15,000	4.16	144,142	15,000	4.16
	Load lightening by SiC of the tunnel kiln	2,508,224	71,444	1.14	2,508,224	71,444	1.14

(continued)

Table 11.1 (continued)

		PMF			RMF		
		Savings	Investment	Pay Back	Savings	Investment	Pay Back
		kWh/year	€	years	kWh/year	€	Years
	Load lightening by SiC of the shuttle kiln	409,046	17,216	1.68	409,046	17,216	1.68
	Heating recovery from compressor refrigeration	569,600	38,400	2.70	569,600	38,400	2.70
Exclusive thermal measures	Heating recovery for mold dryers	2,004,792	216,034	4.31	–	–	–
	Heating recovery for boilers	581,856	62,700	4.31	1,700,102	374,929	8.82
	Heating recovery for heating units	4,351,851	468,949	4.31	609,956	134,515	8.82
	Heating recovery for ware dryers	2,112,526	227,643	4.31	2,112,526	465,882	8.82
	Induced draught for tunnel kiln	3,086,298	150,000	1.94	3,086,298	150,000	1.94
	Primary air preheating for tunnel kiln	3,594,441	200,000	2.23	3,594,441	200,000	2.23
	Primary air preheating for shuttle kiln	1,931,509	150,000	3.11	1,931,509	150,000	3.11
	Photovoltaic plant	2,748,927	809,700	2.95	2,250,072	784,649	3.49
	Cogeneration	7,669,258	2,782,582	3.63	6,277,498	2,823,388	4.50
	Total	34,473,362	8,059,326	4.07	26,885,820	7,254,408	4.72

where:

Inv: total present value of the investment (€)

CRF: capital recovery factor (dimensionless)

Table 11.2 Water saving investments from a pay-back perspective. (*Source* Data compiled by the authors)

		PMF			RMF		
		Savings	Investment	Pay Back	Savings	Investment	Pay Back
		m^3/year	€	years	m^3/year	€	Years
Water-saving measures	Condensation by absorption machine	9,426	943,685	66.74	2,233	223,545	66.74
	MED distillation	13,020	140,298	7.18	89,521	964,664	7.18

Table 11.3 Financial parameters for the present value analysis. (*Source* Data compiled by the authors)

financial parameters		
O and M equipment cost percentage of investment per year	1	%
Effective discount rate	10	%
Economic lifetime (years)	15	years
Inflation rate	3	%
Escalation rate O and M	1	%
Escalation rate fuel cost	2.5	%
CRF	13.15	%
Nominal escalation rate O and M	4.03	%
Nominal escalation rate fuel	5.58	%
$k_{O\&M}$	0.95	
k_{fuel}	0.96	
Constant escalation levelization factor O and M ($CELF_{O\&M}$)	1.3	
Constant escalation levelization factor fuel ($CELF_{\text{fuel}}$)	1.44	
Natural gas price	0.025	€/kWh
Electricity price	0.1	€/kWh
Water price	1.5	€/m^3
Levelized price of natural gas	0.036	€/kWh
Levelized price of electricity	1.44	€/kWh
Levelized price of water	2.164	€/m^3

i_{eff}: effective discount rate (dimensionless)

n: economic life (years).

The levelized cost $LC_{O\&M}$, LC_{fuel} are calculated assuming a specific characteristic increase of the price, which is represented by

$$LC_{O\&M} = C_0^{O\&M} \cdot CELF_{O\&M} = C_0^{O\&M} \cdot \frac{k_{O\&M} \cdot \left(1 - k_{O\&M}^n\right)}{1 - k_{O\&M}^n} \cdot CRF \quad (11.8)$$

$$LC_{O\mathrm{fuel}} = C_0^{\mathrm{fuel}} \cdot CELF_{\mathrm{fuel}} = C_0^{\mathrm{fuel}} \cdot \frac{k_{\mathrm{fuel}} \cdot \left(1 - k_{\mathrm{fuel}}^n\right)}{1 - k_{\mathrm{fuel}}^n} \cdot CRF. \quad (11.9)$$

CELF is the constant escalation levelization factor, which depends on the effective annual cost of the money and the nominal escalation rate, r_n, by means of the parameter k defined in

$$k = \frac{1 + r_n}{1 + i_{\mathrm{eff}}} \quad (11.10)$$

r_n is the nominal escalation rate, which represents the annual change in cost and includes the effects of both the real escalation rate r_r and the inflation r_i.

After these calculations, a new perspective of the analyzed investments is obtained as shown in Tables 11.4 and 11.5.

Table 11.4 Summary of investments to optimize a factory from a LCOE perspective

		PMF			RMF		
		Savings	Investment	LCOE	Savings	Investment	LCOE
		kWh/year	€	€/kWh	kWh/year	€	€/kWh
General electrical measures	Savings by motor efficiency	201,922	1,417,710	1.0280	141,303	992,096	1.0280
	Savings by frequency converters	1,609,294	1,035,251	0.0942	1,273,461	819,211	0.0942
	Savings by LED lighting	79,680	2,622	0.0048	63,410	2,087	0.0048
	Savings by lighting regulation	30,886	21,896	0.1038	24,639	17,468	0.1038
General thermal measures	Increase of thermal insulation of the mold dryer	138,981	15,840	0.0167	–	–	–
	Increase of thermal insulation of the casting hall	255,100	308,040	0.1768	87,786	184,824	0.3083

(continued)

Table 11.4 (continued)

		PMF			RMF		
		Savings	Investment	LCOE	Savings	Investment	LCOE
		kWh/year	€	€/kWh	kWh/year	€	€/kWh
	Thermal and humidity control of the HVU's	343,221	35,000	0.0149	–	–	–
	Increase of thermal insulation of the ware dryer	101,809	13,300	0.0191	101,809	13,300	0.0191
	Thermal and humidity control of the ware dryer	144,142	15,000	0.0152	144,142	15,000	0.0152
	Load lightening by SiC of the tunnel kiln	2,508,224	71,444	0.0042	2,508,224	71,444	0.0042
	Load lightening by SiC of the shuttle kiln	409,046	17,216	0.0062	409,046	17,216	0.0062
	Heating recovery from compressor refrigeration	569,600	38,400	0.0099	569,600	38,400	0.0099
Exclusive thermal measures	Heating recovery for mold dryers	2,004,792	216,034	0.0158	–	–	–
	Heating recovery for boilers	581,856	62,700	0.0158	1,700,102	374,929	0.0323
	Heating recovery for heating units	4,351,851	468,949	0.0158	609,956	134,515	0.0323
	Heating recovery for ware dryers	2,112,526	227,643	0.0158	2,112,526	465,882	0.0323

(continued)

Table 11.4 (continued)

		PMF			RMF		
		Savings	Investment	LCOE	Savings	Investment	LCOE
		kWh/year	€	€/kWh	kWh/year	€	€/kWh
	Induced draught for tunnel kiln	3,086,298	150,000	0.0071	3,086,298	150,000	0.0071
	Primary air preheating for tunnel kiln	3,594,441	200,000	0.0081	3,594,441	200,000	0.0081
	Primary air preheating for shuttle kiln	1,931,509	150,000	0.0114	1,931,509	150,000	0.0114
	Photovoltaic plant	2,748,927	809,700	0.0431	2,250,072	784,649	0.0511
	Cogeneration	7,669,258	2,782,582	0.1562	6,277,498	2,823,388	0.1689
	Total	34,473,362	8,059,326	0.0572	26,885,820	7,254,408	0.0636

Table 11.5 Water saving investments from a LCOE perspective. (*Source* Data compiled by the authors)

		PMF			RMF		
		Savings	Investment	LCOE	Savings	Investment	LCOE
		m^3/year	€	€/kWh	m^3/year	€	€/kWh
Water-saving measures	Condensation by absorption machine	9,426	943,685	2.10	2,233	223,545	3.56
	MED distillation	13,020	140,298	0.45	89,521	964,664	0.57

The resultant LCOE for each investment represents the levelized cost of saving these amounts of energies (water) during the economic life of 15 years. Thus, this cost must be lower than the levelized actual price of this energy (water) to make interesting the investment.

This approach is more accurate than pay-back. However, there is some uncertainty in most of the parameters assumed since the length of the economic period of 15 years. It is a tough task to estimate escalation rates or effective discount rate for such a long time. In any case, this methodology provides a very appropriated tool to analyze the viability of the different investments in order to take the right decision.

References

1. Montes MJ, Rovira A, Muñoz M, Martínez-Val JM (2011) Performance analysis of an integrated solar combined cycle using direct steam generation in parabolic trough collectors. Appl Energy 88(9):3228–3238. https://doi.org/10.1016/j.apenergy.2011.03.038
2. Statistics Eurostat Electricity statistic (2021) https://ec.europa.eu/eurostat/statistics-explained/index.php?title=File:Development_of_electricity_prices_for_non-household_consumers,_EU-27,_2008-2020_(EUR_per_kWh)_v4.png. Accessed 13 Jan 2021
3. Statistics| Eurostat Natural gas (2021) https://ec.europa.eu/eurostat/databrowser/view/nrg_pc_203/default/line?lang=en. Accessed 13 Jan 2021
4. Kanakoudis V, Tsitsifli S (2016) Socially fair domestic water pricing: who is going to pay for the non-revenue water? Desalin Water Treat 57:11599–11609. https://doi.org/10.1080/19443994.2015.1067169

The manufacturer's authorised representative in the EU is Springer
Nature Customer Service Centre GmbH, Europaplatz 3, 69115 Heidelberg,
Germany. If you have any concerns regarding our products, please
contact ProductSafety@springernature.com

Printed and bound by CPI Group (UK) Ltd, Croydon, CR0 4YY
30/09/2026
02225462-0012